# GODS, STARS, and COMPUTERS

# GODS, STARS, and COMPUTERS

*Fact and Fancy in Myth and Science*

## MALCOLM E. WEISS

DOUBLEDAY & COMPANY, INC.

GARDEN CITY, NEW YORK

Library of Congress Cataloging in Publication Data
Weiss, Malcolm E    Gods, stars, and computers.

Includes index.
SUMMARY: Compares ancient myths relating to natural phenomena with modern scientific explanations of the same phenomena.
1. Science — Juvenile literature.   2. Myth — Juvenile literature.
[1. Science.   2. Mythology]
I. Title.    Q163.W42    500

ISBN: 0-385-12488-0 Trade
0-385-12489-9 Prebound
Library of Congress Catalog Card Number 77-80920
Copyright © 1980 by Malcolm E. Weiss
Printed in the United States of America
All Rights Reserved
First Edition

J500
W436G

# Contents

| | |
|---|---|
| *Chapter One:* Tales of Two Mountains | 11 |
| *Chapter Two:* More Volcano Tales | 19 |
| *Chapter Three:* Earthquake Tales | 35 |
| *Chapter Four:* Tales of the Sky | 53 |
| *Chapter Five:* Medicine Has Roots | 71 |
| *Chapter Six:* Animals, Men, and Nightmares | 97 |
| *Chapter Seven:* From Nightmare to Waking Dream | 113 |
| *Index* | 121 |

# GODS, STARS, and COMPUTERS

## Chapter One

# Tales of Two Mountains

Llao, god of the Underworld, sat on his throne atop Mount Mazama. He turned south to glare upon Skell, god of the Overworld. Skell sat atop Mount Shasta, 100 miles away, and glared back.

Then there was war between them.

Smoke and flames rose from their thrones. They hurled red-hot rocks and balls of fire at each other and roared their battle cries to the skies.

The struggle went on and on. By day the smoke of their warfare hid the sun. By night the only light came from the red flames of their anger. The land itself seemed ablaze.

At last Skell gained the victory. With one mighty final roar, the top of Mount Mazama exploded and caved in on itself. Llao fell through the collapsing mountaintop back down into the Underworld. Where his throne once stood, there was an enormous bowl-shaped crater. Skell caused the crater to be filled with rainwater as a sign of peace.

This is a story the Klamath Indians of the American Northwest tell about how Crater Lake, in Oregon, was made. Here's another tale about how Crater Lake came to be:

Mount Mazama was an active volcano. During a particularly violent eruption, Mazama literally blew its top. The sky was darkened. The peak of the mountain collapsed inward on itself. Over the centuries, the crater slowly filled with rainwater.

Leave out the gods, and the two stories, one mythical, one scientific, are practically the same. Why are they so much alike?

*Opposite*

The rim of Crater Lake. When Mount Mazama caved in about 7,000 years ago, it left a hole 6 miles across and perhaps 2,000 feet deep. The hole slowly filled with rainwater and formed Crater Lake. The steep shoreline shows how the mountaintop fell in on itself. (*United States Department of the Interior, National Park Service Photo*)

A winter scene along Crater Lake. (*United States Department of the Interior, National Park Service Photo*)

The scientific story is based partly on observations of how other volcanic craters similar to Crater Lake have been formed. In addition, there is clear evidence of Mount Mazama's active past. Layers of volcanic ash lie in the soil all around the mountain. Analyzing these layers, scientists have found that Mazama blew up around 7,000 years ago.

There are other clues in the volcanic ash, clues that help answer our question. Buried in the ash are Klamath Indian sandals, pottery, tools, and weapons. So the Indians were living there when the volcano erupted.

Mount Shasta, in California, 100 miles south of Crater Lake, has been an active volcano, too. The story of Skell and Llao seems to show that both volcanos were active at the same time.

To the Klamath, this tremendous violence could be explained only as warfare. The whole series of eruptions was a titanic battle between two superbeings. Within the limits of their knowledge, the struggle of these gods was a reasonable explanation through storytelling, based partly on fact and *partly on imagination.*

So is the scientific explanation. This may seem like an outlandish idea. But it is not. What *is* outlandish are the kinds of things we usually think of when we hear the word "science."

Science is true. Science is certain. Scientists are detached, objective observers. They follow in Mother Nature's footsteps murmuring, "Just give us the facts, ma'am." And from the facts, they arrive, step by logical step, at clear and precise theories.

This might be called the "Madison Avenue view" of science. Only on such things as TV commercials for headache remedies do "scientists" in white lab coats dole out Truth and Certainty.

This view did not start with Madison Avenue, however. Real scientists, although they know better, sometimes encourage people to think like this about science. Indeed, scientists come under the spell of Science as Final Truth themselves, once in a while. For example, the biologist Dr. A. Bentley Glass said:

"The great conceptions, the fundamental mechanisms, and the basic laws are now known. For all time to come, these have been

The Phantom Ship, a small island in Crater Lake. After the collapse of Mount Mazama, small volcanic cones formed on the floor of what was to become Crater Lake. These cones were the last gasp of the eruption. The Phantom Ship, now tree-covered, is one of them. (*United States Department of the Interior, National Park Service Photo, George A. Grant*)

discovered, here and now, in our own lifetime . . . We are like the explorers of a great continent who have penetrated to its margins in most points of the compass and have mapped the major mountain chains and rivers. There are still innumerable details to fill in, but the endless horizons no longer exist."

Glass, a distinguished scientist, made those remarks in 1970, when he retired as president of the American Association for the Advancement of Science. Glass's opinions had barely been heard before new discoveries by his fellow scientists were proving him wrong.

Seventy years earlier, another famous scientist, Lord Kelvin, had said much the same thing about physics. Even as he spoke, the basic laws of physics were about to be rewritten by scientists like Albert Einstein.

Rewriting those laws took hard work and hard thinking. But there was no logical, step-by-step road to the new ideas. There was, instead, the leap into the dark beyond the facts. There was the sudden flash of inspiration. "There is no logical path to these laws . . . Only intuition . . . can reach them," wrote Einstein.

And he went on: "Man tries to make for himself in the fashion that suits him best a simplified . . . picture of the world; he then tries to substitute this cosmos of his for the world of experience and thus to overcome it. This is what the painter, the poet, the speculative philosopher, and the scientist do, each in his own fashion."

This is what men have always done. Some sixteen thousand years ago, the animals slain in the hunt came to life again. Some unknown artists painted the bear, the bison, the boar, the rhinoceros, and the woolly mammoth on cave walls where they can still be seen and admired today.

The paintings are often far back in the caves. They are in pits, or at the ends of narrow winding tunnels, or high up on the cave walls.

The artist tried to draw as complete a picture as possible. He (or she) drew the eyes, tusks, trunk, shaggy body, and stumpy tail of the woolly mammoth in detail. It was a way of making a picture of some part of the world, of freezing in time the experience of the hunt.

For thousands of years, man painted his hopes and dreams on the walls of his caves. Because the animals hunted were an important part of those hopes, he drew many of them. The menagerie in his mind's eye came to life on the bare stone walls.

Over the years the pictures changed. The woolly mammoth was no longer painted in detail. The eyes became simple circles; the feet, four ovals not even attached to the body; and the shaggy coat, a fuzzy, blurry bunch of roughly parallel lines. It was less a complete picture of a mammoth than a picture of the general *idea* of a mammoth — something massive, tusky, shaggy, and slow.

The painter, as Einstein said, had made "a *simplified* picture of the world."

Even the lines weren't necessary. One word could express the entire idea: "mammoth." The word is a picture of the idea in sound. Words could be put together to tell the complete story of a mammoth hunt.

The mind of early man gave shape to ideas in other ways. The hand that turned a stone into an ax blade was guided by an idea — an idea of what the stone could become.

Ideas changed. New ideas were born. Tools and weapons became sharper, and the handles fit the hand more comfortably. More than that, the tools and weapons were beautifully decorated. And fine lines were chiseled on mammoth tusks in regular patterns like the patterns of lines on a ruler.

The lines on a ruler stand for numbers. They may be numbers of inches, centimeters, or other units. The lines on the tusks stand for numbers also. They make regular patterns that seem to match the changing shape of the moon through the month. To see that pattern must have taken months of careful moon watching.

Painting, language, storytelling, and numbers — they all grew out of the same idea-shaping power of the mind. They are all made from the same mixture of careful thought, imagination, and joy in both.

In the shaping of the tools and the number patterns carved on some of them are the beginnings of science. In the decorations are the beginnings of art.

Today, the artist, the writer, and the painter seem far apart from the scientist. But they have much in common. This look at old tales of wonder and new tales of science explores that common ground.

## Chapter Two

# More Volcano Tales

The Isle of Skye is rugged and beautiful. It is the largest of the Inner Hebrides, just northwest of mainland Scotland. On the southwest coast of Skye, a long range of steep, jagged hills rises sharply from the shore. The tallest of the hills is about 3,300 feet high.

These hills, the Cuillins, were formed by the Sun. Or so goes the local tale. Cailleach Bhur, the Hag of the Ridges, captured a maiden on Skye long ago. The girl begged the Sun to help her. The Sun hurled a flaming spear at the Hag, but it missed and stuck into the ground. A dome-shaped blister formed. The blister swelled and swelled. It burst and a fiery liquid poured out. The frightened Hag ran away and never returned. The red-hot liquid spread and hardened to form the Cuillins.

Geologists, on the other hand, say the Cuillins are made of gabbro. Gabbro is a kind of rock, usually dark-colored and made up of many coarse grains or crystals. The Palisades along the western shore of the Hudson River in the northeastern United States are a form of gabbro.

Where does gabbro form? It begins as molten rock, miles beneath the surface of the earth. This subterranean molten rock is called "magma." Magma is a mixture of liquid rock and various gases, among them steam, ammonia, and hydrogen sulfide. The magma is very hot — from 1,000 to 2,000 degrees Fahrenheit (from 540 to 1,100 degrees Celsius). Under high pressure, deep below ground, the gases are held in solution in the liquid rock, just as carbon dioxide is dissolved in an unopened bottle of soda water.

Magma is lighter than the surrounding solid rocks when it rises through the earth. As it nears the surface, the overhead pressure decreases and the dissolved gases begin to seethe out. Pushing aside or melting the rocks above, the magma may burst through cracks and vents in the earth's surface as lava. Exposed to air — or water, if the vent lies on the ocean floor — the lava cools quickly and solidifies. The hardening lava may build up into a cone-shaped hill or mountain with one or more vents at the top and along the sides — it has formed a volcano.

The volcanic rocks themselves hold the clues to how they were made. When a liquid solidifies very quickly, crystals have no chance to form and the resulting solid is glassy. Obsidian is a glassy volcanic rock formed when lava cools very rapidly. Lava that cools a little more slowly will have very fine crystal grains in it. In fact, the size of the grains in rocks formed from magma or lava is a direct key to how fast the molten rock cooled. The bigger the grains, the more slowly the rock cooled.

As we've seen, gabbro is very coarse-grained. That means it cooled very slowly — much more slowly than would be possible if it had hardened above ground.

Gabbro and other similar rocks come from magma that solidifies underground. The Cuillins were made in this way when magma pushed the overlying rocks into a dome shape and then hardened. Once the Cuillins were the very foundation stones of mighty volcanoes. Time passed and the volcanoes died. Wind and water wore them down to the ground. The ground itself was worn away, exposing the Cuillins like the foundations of a long-vanished building.

But that brings us back to the tale of the Hag and the Sun. For the image of a dome, swelling and swelling and spewing out a fiery liquid that hardened, sounds almost like an eyewitness description of what really took place.

Except that the Cuillins formed many millions of years ago, when there were no people on Skye — or anywhere else, for that matter.

And except that the Cuillins formed miles beneath the ground.

Perhaps Skye was settled long ago by people who came from a

place where there *were* active volcanoes. But it seems as if there is an easier explanation for the story.

The Hag of the Ridges also stands for winter. The maiden she captured symbolizes spring. Spring naturally calls on the Sun for help. Of course, a spear from the Sun might be expected to melt rock.

And what about the Hag fleeing and never returning? Well, that stands for the fact that snow never stays on the tops of the Cuillins because they are so steep.

The match between this legend and the story science tells appears to be merely coincidence. But let's take a look at still another volcano tale . . .

The most famous volcano legend of all is the Hawaiian legend of Pele, the goddess of volcanic activity. The Hawaiian Islands were settled about a thousand years ago by the Polynesians. These expert sailors crossed the vast Pacific Ocean, navigating from island to island in their long canoes.

According to the Polynesians, Pele came to the Hawaiian Islands from some other part of the Pacific. Her older sister, Namakaokahai, had driven her from their home after a fierce fight. Pele first tried to make her home on the island of Kauai.

Pele's idea of home was the heart of a volcano. To build herself one, she used her magic digging tool, Paoa. Wherever she thrust Paoa into the ground, a volcanic crater formed.

Pele dug furiously on Kauai. The dirt she hurled up formed a large hill, Pele's Hill, which can still be seen. But before Pele could finish, her sister caught up with her and killed her.

But Pele came back to life and moved on to Oahu, the next island in the Hawaiian Islands chain. At Diamond Head, she dug until she struck fire. Home at last!

But it was not to be. Water flooded in and put the home fires out. Pele moved on to Molokai and the same thing happened. Next came Maui and there she built the volcano Haleakala. Soon smoke was pouring from Pele's chimney atop Haleakala.

Lava from Haleakala on Maui Island formed this rugged black coastline. (*United States Department of the Interior, National Park Service Photo*)

MORE VOLCANO TALES

Namakaokahai saw the smoke and once again pursued Pele. They fought and again Pele was killed.

Namakaokahai set sail for home. She looked back over the waves to enjoy her triumph one more time. And there, over the last and biggest island, Hawaii, smoke and flame billowed high into the air. Namakaokahai saw that her sister's spirit could never be killed and she left her in peace.

We don't know how long Pele's journey across the Hawaiian Islands took. But since she started excavating on Kauai, clearly the craters on that island must be the oldest. Oahu's are next oldest, then Molokai's, then Maui's, and finally Hawaii's.

And this is exactly what the scientific story of the islands tells us. Pele's Hill, on Kauai, is the cone of an extinct crater, millions of years old. It is so old that wind and rain have worn it down to a hill. Diamond Head, on Oahu, is all that remains of a volcano that was active about 150,000 years ago. Haleakala last erupted in the late 1700s. On Hawaii, the biggest island, there are now two active volcanoes: Mauna Loa, the world's largest, and Kilauea. A third, Hualalai, last erupted about 1800.

Pele's pit of fire at Diamond Head was flooded out. Again, the scientific story of the islands tells us much the same thing. The volcano that formed where the crater of Diamond Head now is thrust up near the shore. It was soon worn away by the action of the sea.

When the Polynesians arrived in the islands about a thousand years ago, there were thus three active volcanoes on Hawaii and one, Haleakala, on Maui. But the legend of Pele seems to reveal a knowledge of the volcanic history of the Islands that goes much further back in time. The match between Pele's story and the scientific description of how the Hawaiian Islands formed is too good to be mere coincidence.

Long before they came to Hawaii, the Polynesians knew about volcanoes. The Pacific Ocean is dotted with volcanic islands. They had seen volcanoes of many kinds. They had seen what happens when the sea inundates a volcano. They had seen how beaches and

The main crater of Haleakala, last active in the 1700s. The dark band across the middle of the photo marks an old lava flow. Smaller young craters lie along the floor of the main crater. (*United States Department of the Interior, National Park Service Photo*)

Lava plunges into the Pacific Ocean along the shores of Hawaii, creating clouds of steam. The Polynesians must have seen many such battles between fire and water on their journeys across the Pacific. (*United States Department of the Interior, Geological Survey Photo*)

Lava and rock fragments spew skyward from Kilauea, one of the two active volcanoes on the island of Hawaii. (*United States Department of the Interior, Geological Survey Photo*)

islands can be worn away by the waves. They knew that lava flows destroy plants and animals in their path, but that slowly the vegetation comes back to cover old flows once again. And these things were passed on from generation to generation in stories and sayings.

They used what they knew, perhaps without even thinking about it, when they created the legend of Pele. Hawaii is the tallest of the islands, the twin peaks of Mauna Loa, snow-covered in winter, towering 13,700 feet. On the southeast shore, where volcanoes are still active, the beaches are black sand — glasslike bits of new lava. In fact, the island is still growing as new lava adds to its surface.

Maui is older and not so high. The peak of Haleakala is 10,000 feet above the sea. Beaches of coral sand have replaced the lava flows of long ago. Molokai, Oahu, and Kauai are much more worn down, none more than 5,000 feet high. They are covered with vegetation, and a variety of animals live on them.

All these clues gave the Polynesians an idea of the order in which the Hawaiian volcanoes were formed. To explain *how* they were formed, they invented Pele. But Pele, goddess though she was, worked *within the order the Polynesians read from the appearance of the islands*.

Scientists today know much more about the structure of the earth and what goes on inside it. They do not need Pele to explain how the Hawaiian Islands formed.

Scientists now think that the earth's crust of surface rocks is broken up into a number of plates. This crust of rocks is about 20 miles thick, but compared to the size of the earth, it is about as thick as the skin of an apple.

The boundaries of the plates run along the ocean floor for the most part. If we could drain the oceans and look at the plates and their boundaries from a long way off in space, the earth would look a bit like an egg shell with cracks running around it.

Like gigantic rafts, these plates "float" on heavier rocks that lie beneath them. The plates change position very slowly, carrying the continents and islands with them. But this is an extremely slow change. It takes many millions of years to become noticeable.

Most of the boundaries of earth's plates lie hidden beneath the waves. They form a system of rifts in the sea floor — a system some 45,000 miles long. But there are places where the sea floor comes ashore. The island of Iceland, created by volcanic action, lies astride the Mid-Atlantic rift. Here we see a part of that rift made visible as it cuts through Iceland. This dry-land portion of the rift system is called the Almannagjá cleft. (*Icelandic National Tourist Office Photo*)

The ramparts of the Almannagjá cleft at Thingvellir, Iceland. (*Icelandic National Tourist Office Photo*)

The Hawaiian Islands lie somewhat north of the center of one vast plate — the Pacific plate — which contains most of the Pacific Ocean. The plate is shaped roughly like Africa. though it's three times Africa's size.

The Pacific plate is slowly turning. As it turns, the Hawaiian Island chain is imperceptibly drifting in a northwesterly direction.

Right now, the island of Hawaii is over what geologists call a "hot spot" in the earth's interior, below the crust. At this spot, magma is driven upward toward the surface. It is this magma that built the island. Hawaii itself is the top of a volcanic mountain rising some five miles from the floor of the sea.

Millions of years ago, Maui lay over the hot spot. Or rather, the place in the ocean where Maui is now lay over it. And so, as the plate has turned, a succession of volcanic islands have been born, rising from the waves, steaming and belching lava. In time, the volcanoes of the island of Hawaii will grow quiet as it drifts away from the hot spot and a new island will be born southeast of it.

With their new knowledge, the geologists can trace the Hawaiian chain still further back in time — and space. The Hawaiian Islands extend some 1,600 miles to the northwest from Hawaii. The oldest islands, at the northwestern end of the chain are low and sandy, worn down almost to sea level — Laysan, Lisianski, Midway, and Kure. Along this stretch of the Pacific, too, lie dangerous reefs that are hazards to any ship. These are the barely submerged remnants of still other parts of the chain.

Up to the early 1960s, most geologists did not accept the idea that continents and islands moved. They did not know enough then about the interior of the earth to explain how continents *could* move.

But scientists did know that continents far apart — like Africa and South America — had many similar kinds of plants and animals millions of years ago. And that posed the question: How did they get from one continent to the other? To answer this question, scientists came up with the idea of a bridge of land between the two continents. There was not much evidence that such a bridge ever existed — except for the fact that the two continents had similar forms of life.

The trees are seared and the land charred for miles around in the path of lava from Kilauea. (*United States Department of the Interior, Geological Survey Photo*)

An aerial view of Kilauea in action. A lake of lava fills the volcano's crater to the brim and streams over the edges in a fiery flood. (*United States Department of the Interior, Geological Survey Photo*)

But the idea of land bridges seemed the only reasonable explanation. A few scientists suggested the idea of continental drift — but they, in turn, could not give any convincing explanation of what it was that made continents move.

After World War II, new evidence came in. Echo-sounding devices and other instruments made it possible to explore the ocean floor in detail. The great boundaries of the plates were mapped completely.

These discoveries should have made continental drift convincing, according to geologist Dr. Bruce Heezen. But, he pointed out, more than facts were involved. Scientists were split into "drifters" and "antidrifters," as if they were members of two opposing religions. Antidrifters complained the new theory would mean geology would have to be rewritten. "The majority," Heezen wrote, "were clearly not willing to grant that they had been so greatly in error."

Finally, though, the evidence became overwhelming. Today, continental drift has become respectable. Still, as Heezen has noted, the name "continental drift" stirs up many bad memories of an old scientific feud. So many scientists prefer to call the theory by another name — such as "continental displacement."

The names don't matter. What's important is that scientists now believe that South America and Africa were once joined together, as were other continents. The invention of the land bridge, like the invention of Pele, is no longer needed. What was once part of a scientific story has gone the way of a myth.

The idea that has replaced it was built up from a large number of facts. But what holds the facts together is a daring and awesome act of imagination — for no human eyes have ever been witness to the creation of the Hawaiian Islands or to the titanic forces that stir the earth's interior.

It is a vision of islands turning on a turntable as wide as the moon — a turntable driven slowly but irresistibly by the central fires of the earth.

It is a vision the old Polynesians, who dreamed of many things on earth and in the sky, would have appreciated.

*Chapter Three*

# Earthquake Tales

The vision of continents drifting is an awe-inspiring one. But it is a vision for the mind's eye alone. In the whole time that man has existed on this planet, the motions of the continents have scarcely been noticeable.

But these slow, immensely powerful movements can cause other changes — changes that take place with devastating speed. As one of the earth's plates grinds against another, rocks on the surface and beneath it are strained. The strain may build up for years or centuries. Then whole sections of rock beneath the surface break and slide past each other.

At 3:42 on the morning of July 28, 1976, such a break occurred some miles below the Chinese city of Tangshan. A few moments before, weird white and red lights lit up the sky. They could be seen 200 miles away. People were awakened by the glare, thinking their room lights had accidentally turned on.

The ground twisted and heaved violently. The jolt hurled people to the ceiling. Buildings crashed in ruins, thousands of them falling down, one after another, like a pack of cards. Trees snapped like toothpicks, railroad tracks were tangled and twisted out of shape. Half a mile from the earthquake center, an entire field of corn was knocked flat, the fallen plants all pointing away from the city.

A second huge shock occurred sixteen hours later. In all, about 700,000 people were killed. It was the second worst earthquake in history. The worst quake on record was in the year 1556, also in China. That one resulted in 830,000 deaths.

All major earthquakes begin with this massive breaking and sliding of rocks underground. Shock waves travel in every direction through the earth from the point of breakage — the focus of the earthquake. The place on the surface closest to the focus, the epicenter, is the place the shock waves reach first. There the waves are strongest and do the most damage.

But all around the world, shock waves reaching the surface are detected by seismographs. The seismograph measures the strength of the waves and records the time of their arrival. Some of these waves have passed through the upper mantle of the earth's crust. Some have skirted the edge of earth's core 2,000 miles down. Other waves, reaching the surface of the earth at the place opposite the epicenter, have traveled through the center of the earth.

The waves are of two kinds: P-waves, also called push, or pressure, waves; and S-waves, also called shake, or shear, waves. P-waves travel twice as fast as S-waves and can move through solids, liquids, and gases. S-waves can only move through solids.

From the way the waves behave on their journey through the earth, seismologists have built a model of earth's structure. There is the crust; the slowly flowing rocks of the upper and lower mantles; the liquid outer core, through which S-waves won't go; and the solid inner core. Thus these destructive waves have served at least one useful scientific purpose. They have "taken a picture" of the interior of the earth.

The dangers of earthquakes are of increasing concern to scientists everywhere. Parts of China are earthquake-prone. So, for example, are Japan and California. The California-Nevada region has averaged about five thousand quakes a year since the first recorded quake in 1769. Most of these have been minor, but a few, particularly along the 600-mile-long San Andreas fault, have been major ones. The San Francisco quake of 1906 resulted in terrible damage, largely due to the three-day city-wide fire that followed. What's less well known is that places like Boston, Massachusetts, and Kansas City, Missouri, lie in earthquake-active zones. In fact, the 1811 Missouri quake was the largest on record in North America.

# EARTHQUAKE TALES

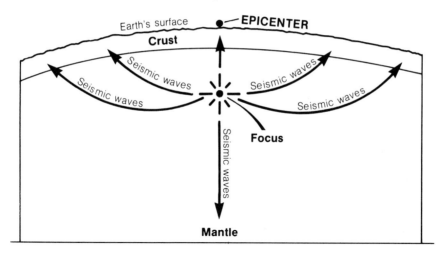

Waves generated by an earthquake help pinpoint its focus in the earth. The quake's epicenter lies directly over its focus.

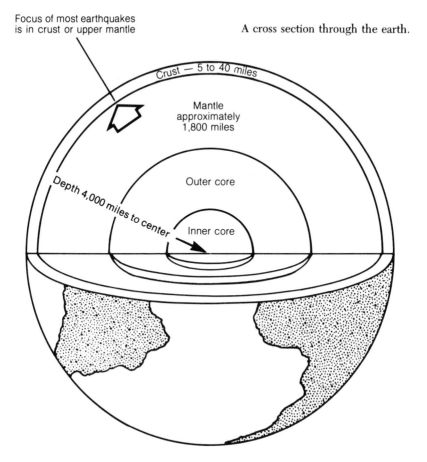

A cross section through the earth.

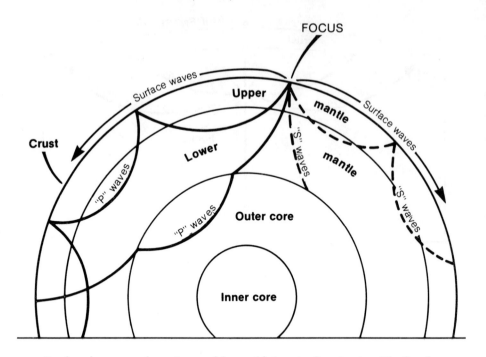

Earthquake waves take a picture of the earth's interior for scientists. The fact that S-waves cannot pass through the outer core shows that it is liquid.

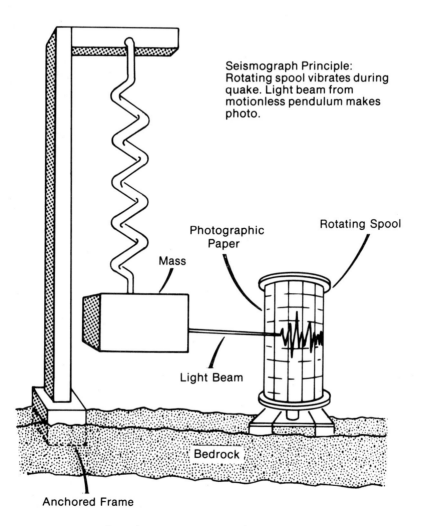

How a seismograph works.

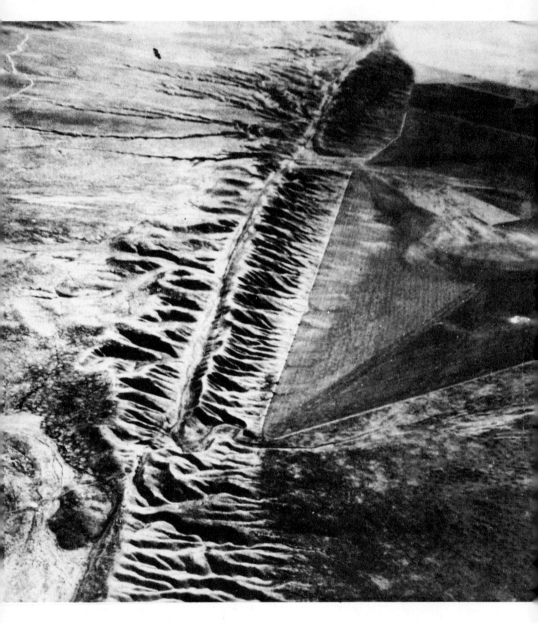

An aerial view of the San Andreas fault in the Carrizo Plain area of central California. The fault is some 600 miles long and cuts into the earth to a depth of over 20 miles. (*United States Department of the Interior, Geological Survey Photo, J. R. Balsey*)

Missouri didn't have many people at the time, so few lives were lost and not much attention paid to the quake. But since 1811 the population of the world has soared. Cities are crowded. Power plants, factories, roads, and farms are widespread. All this makes it more likely that a major earthquake almost anywhere could be devastating, like the one at Tangshan.

The danger can be lessened if we design earthquake-resistant buildings and roadways. In earthquake-prone areas, strict building codes can be passed.

There is another possibility that could save many lives and even prevent some damage: earthquake prediction. If earthquakes could be predicted accurately some weeks ahead of time, people could be evacuated from dangerous areas. They could live in temporary shelters — tents and shacks — in the countryside until the quake was over. Patients in hospitals could be transferred from the

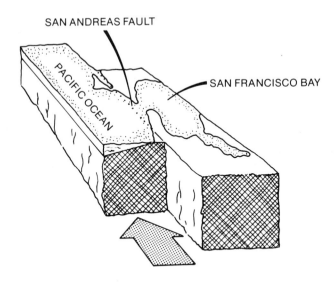

The San Andreas fault marks the boundary between two of the earth's plates. The rotation of the Pacific plate carries the land west of the fault line slowly northward at a rate of about 2 inches a year.

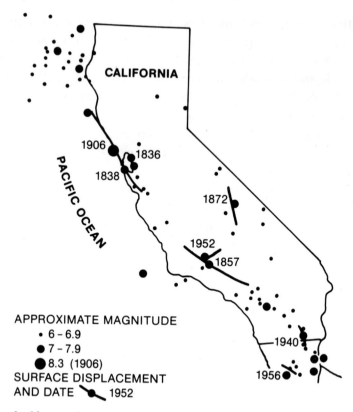

As strain builds up in the rocks lying across a fault line, they fracture periodically. In California, the epicenters of resulting earthquakes lie largely across the San Andreas fault and other faults branching from it. The location of these faults is shown on the map by the surface displacement lines. The dots show the epicenters of earthquakes in the area. The size of the dots is a measure of the magnitude of the quake on the Richter scale. A magnitude of 2 indicates a weak quake that can barely be felt. A magnitude of 3 indicates a quake releasing about sixty times as much energy as a quake of magnitude 2, and so on for each unit increase on the scale. Magnitudes of 7 or higher indicate a severe quake causing major destruction.

earthquake zone. Gas and oil pipelines could be shut down. Traffic could be detoured. Medical and rescue teams could be alerted.

In fact, this has been done in China. In February 1975 a severe quake struck Haicheng, 300 miles northeast of Tangshan. Many

The network of faults in coastal California.

buildings were destroyed, but there were very few casualties. Two months earlier the Haicheng area got its first general warning that an earthquake was likely. Planning for the quake began. At the end of January there was an "earthquake-imminent" warning and the city was evacuated.

The Chinese have set up an earthquake prediction network that involves several hundred thousand people. Workers and farmers scattered around the country check instruments at observation posts several times a day. They keep track of clues that may foretell an earthquake.

A series of small quakes may precede the big one by a few weeks or longer, for example. Among other clues are small magnetic and electrical changes in the rocks and sudden changes in the water level in deep wells.

Chinese scientists admit that the prediction program is still a very crude one. Though several quakes have been successfully predicted, there have been false alarms. Another earthquake-imminent warning had been issued for Haicheng in December 1973. Thousands of people were forced to leave their homes in bitterly cold weather. And there was no earthquake.

Worse yet are the failures to predict a quake that *does* occur. In January 1976 — six months before the disastrous Tangshan quake — there was a general warning that a quake was likely. The warning said the center of the quake would be *either* in the Tangshan area or in the area of the captial city, Peking, about 100 miles to the west. But no earthquake-imminent warning was ever issued in July 1976 and there was no evacuation.

Why not? First, there was no pattern of small shocks leading up to the major quake. Secondly, other clues were uncertain or absent. One clue does not mean very much by itself. The more clues there are, the more likely is the prediction to prove correct. And there are probably lots of clues that haven't yet been discovered. As one Chinese scientist put it, "We don't understand exactly all that happens before an earthquake, but we have to try to use everything."

Western scientists have come to the same conclusion. Up to 1960 not much thought was given to the possibilities of predicting earthquakes. Yet long before an earthquake, rocks underground are slowly straining and bending under enormous pressures. There should be some way of detecting it.

"It's like breaking a pencil; the earth cannot break without things happening beforehand," says Dr. Don Anderson, a California seismologist.

But *what* happens? To start finding out, seismologists began going through the records of past earthquakes. They looked for evidence of unusual events in the weeks or months before the quake. They looked for anything that usually seemed to happen before a quake. And they came up with a list of clues like those we've already mentioned.

Searching through old records, though, has its limits. The scientists who kept those records were not looking for ways to predict earthquakes. They were measuring the strength of quakes or using the earthquake waves to form a picture of the earth's interior. So they may easily have overlooked some clues.

Scientists who are now looking for clues to aid in the prediction of earthquakes have a similar problem. They don't know exactly what to expect. There are so many possible things that might be observed before an earthquake. Some, like the color of a colleague's shirt the day before an earthquake, are obviously not clues. Other things that seem to have little to do with quakes may turn out in the end to be clues.

For instance, what about those lights in the sky over Tangshan? Were they really there? Or did people invent them during the terror of the earthquake? The earthquake caused many fires. Did the people of the city see the glare of the flames reflected in the sky *after* the quake? And did they later remember the lights as something that seemed to come *before* the quake? That's the way stories like these sometimes get started.

Stories about weird lights in the sky around the time of an earthquake are very common. During a quake in Japan in the 1930s the sky was said to glow as if lit up by distant lightning streaks. Other stories told of streamers, columns, and fireballs of ghostly colored lights moving through the sky before the quake. During a 1969 quake in Santa Rosa, California, people claimed to see lights like sparks and meteors streaming aloft.

Up to the time of the Tangshan quake, however, few scientists took the talk of earthquake lights very seriously. Few scientists even considered investigating the stories.

Why not? The reason usually given by the scientists themselves is that there was a lack of carefully gathered evidence, evidence arrived at independently by several observers. This explanation seems to go around in a circle. If *no* scientist even investigated the tales, there would probably never be any hard evidence for *or* against them.

One can guess at other reasons why the tales were ignored. Lights in the sky don't seem to have any likely connection with earthquakes. What's more, these lights come in an odd assortment of shapes and colors. It sounds vaguely silly and quite a bit like those far-out stories about "flying saucers." Scientists, like the rest of us, do not enjoy appearing foolish. They are sensitive about investigating things their colleagues may look at askance.

In the long run, though, truth will out. It was inevitable that some scientists would start looking at these tales more carefully.

Not surprisingly, the investigation began in one of the most earthquake-prone nations in the world — Japan. During the late 1960s, a physicist, Dr. Yutaka Yasui, began collecting photographs of earthquake lights. That collection was one piece of fairly hard evidence.

Still, not much attention was paid by Western scientists. Then, a year after the Tangshan disaster, Drs. Cinna and Larissa Lomnitz were invited to visit China. They are a husband-and-wife scientific team from the National University of Mexico. For the first time, Chinese scientists were allowed to talk freely about their earthquake prediction methods and about what had happened at Tangshan. The Lomnitzes reported that many Chinese scientists told them about the variety of lights seen in the sky over the city just before the quake.

The photos and the widespread confirmation by scientists who were eyewitnesses to the event were too much to ignore. Now an American geophysicist, Dr. John S. Derr, has begun a serious investigation. Other scientists have followed suit, and scientific papers about the lights are beginning to be published.

The first question is: If the lights are related to earthquakes, what's the connection? Gases, such as those that make up the air, glow when a strong electric current passes through them. That's what causes the blaze of a streak of lightning and the glow of a neon tube. But what's the connection between such a strong current and earthquakes?

The most promising theory so far is that the electric current is generated in the rocks by increasing strain — the same slow increase that eventually triggers the earthquake itself.

It's been known for about a century that when certain crystals are put under strain they become electrically charged. The greater the strain, the stronger the charges that develop on the surfaces of the crystal. This is called the "piezoelectric effect," from the ancient Greek word *piezein*, meaning "to press."

Crystals of this kind are used in many kinds of electrical equipment, such as loudspeakers, microphones, and phonograph pickups. Quartz is the most common crystal that shows the effect, and quartz is a major ingredient of rocks in the earth's crust and mantle.

These rocks, some seismologists suggest, may generate large electrical charges under the severe strains that precede an earthquake. The charges may become strong enough to produce sky glows before and during the quake itself.

If this theory is correct, these large-scale electrical charges would occur before every earthquake. They could be easily detected. That would make them an invaluable tool in predicting earthquakes reliably — a tool that might save millions of lives in the future, not to mention damage to property.

We've mentioned earlier, on p. 00, that small electrical changes in rocks are already used as clues in earthquake prediction. These changes, however, are harder to detect and keep track of. Moreover, they are not completely reliable as earthquake predictors.

Nevertheless, they *are* electrical changes caused by strain on the rocks. So it may seem strange that scientists did not ask themselves years ago whether there could be a piezoelectric effect in such rocks, too. It seems like an obvious question.

But that's not really true. It only seems obvious now because we're looking back on it. In those days, scientists were concentrating on other effects. They were looking at how earthquakes work from a different point of view. Only when they began to think of sky glows and the large currents that are needed to produce them did the question become obvious.

Generally speaking, that's what discovery is all about. It is the putting together of familiar things in new ways. It is seeing connections where none were thought to exist.

When one unexpected connection falls into place, others often follow in a rush. Now that scientists believe there is really something to be investigated about earthquake lights, they are noticing how common stories about such lights have always been. The tales go back probably into prehistory. Dr. Derr, the man who started serious investigations of the lights in this country, quotes an ancient Japanese *haiku*:

> The earth speaks softly to the mountain
> Which trembles
> And lights the sky.

It might also be added that the search for unexpected and odd connections between things can make for unexpected and odd combinations of scientists.

For instance, what do a biologist, a geophysicist, and a psychiatrist have in common?

The answer: Chimpanzees.

Well, let's put the question another way: What makes fish jump out of water, cows bawl, rabbits bump their heads, chimpanzees stop climbing, and rats run into houses while cats are running out of them?

The answer: an earthquake that's about to happen.

Okay, so the answer to the first question should have been "chimpanzees *and* earthquakes." That gets the geophysicist into the act.

Like the tales of earthquake lights, the tales of odd animal behavior go back to prehistory. And scientists have been very skeptical about these tales, too, until quite recently.

# EARTHQUAKE TALES

The Chinese working in the earthquake prediction program have kept a close watch on unusual animal behavior as another clue to coming quakes. But, as the Lomnitzes point out, even the Chinese scientists in the program "play down the animal studies." Could it be that there's a little embarrassment there, a desire not to seem foolish before their Western colleagues? If so, the Chinese need not worry. Their Western colleagues are beginning to zero in on the idea of animal behavior predicting earthquakes.

The similarities of animal research with what happened in the earthquake-light research are interesting. Again, the scientists involved went back through old records. This time the locale was Stanford University, in earthquake-active California.

Stanford University has an outdoor primate facility where the chimps are kept and their behavior patterns studied. Dr. Bruce Smith, a geologist, suggested the search of the records to Drs. Helena C. Kraemer and Seymour Levine of Stanford.

Smith told them there had been twenty-five minor quakes in the Stanford area between June 19 and June 24, 1976. What, he asked, had their chimps been doing on the days just before? Had they been acting oddly?

The answer was yes. Before two of the strongest quakes, the chimps had been restless. But instead of climbing, they spent much more time on the ground than usual. The changes in behavior, said Kraemer, were "so significant that it seems unlikely they were due to chance."

In the fall of 1976 a group of American scientists met in Menlo Park, California, at the Center for Earthquake Research. They called for a complete investigation of the idea that animal behavior can be used to predict earthquakes. Experiments with colonies of pocket mice in artificial burrows and kangaroo rats in cages above ground began the next spring. The colonies are located just northeast of Los Angeles in a desert area. This is an area that has frequent small quakes. Some scientists believe a major quake may occur there in the near future. Recording equipment keeps track of the animals' behavior throughout the day and night.

This is far from the first of such experiments, however. Perhaps the first scientific look at pre-earthquakes animal antics was at Tohoku University in Sendai, Japan, in 1932.

Dr. Shinkishi Hatai had several tankfuls of catfish on his laboratory table. Water was pumped from the ground through the tanks and out again in a continuous flow.

Ordinarily, a light tapping on the table didn't bother the catfish at all. But six to eight hours before an earthquake, the catfish almost always became quite sensitive to tapping. They jumped and skittered about.

How did they know? What kind of signal were they getting?

Hatai discovered that if the tank water was *not* passed through the ground, the catfish no longer responded to tapping shortly before an earthquake. So the signal was coming in with the water.

Catfish are known to be able to sense very weak electrical fields in the water, as are many other species of fish. So it may be that changes in these fields before an earthquake make the fish more sensitive to other disturbances outside, such as tapping.

American scientists have similar explanations for the behavior of other animals preceding an earthquake. They may be sensitive to small electrical or magnetic changes in the rocks. Or they may actually *hear* the rocks being strained. Rocks strained to the breaking point emit very high-pitched sound waves, far beyond the range of the human ear.

Like his American colleagues, Hatai became interested in the link between animals and earthquakes because of stories he had heard. In Japan there are dozens of stories about catfish foretelling quakes. In one version, a fisherman wise in the ways of catfish notices how restless they are. Instead of going on fishing, he rushes home just in time to rescue his family.

Oddly enough, a very old Japanese legend says that a giant catfish living beneath the earth is the *cause* of earthquakes. His head lies under a part of Japan. When he wiggles — earthquake!

Stories about animals causing earthquakes by moving about underground are common around the world. And perhaps this catfish legend suggested the idea that catfish can predict earthquakes to the Japanese. When the big catfish begins to get restless, so do the little ones.

But it seems more likely to have happened the other way around. In the past, people lived much closer to animals than they do today. They raised animals for food, used them for work, and learned the ways of wild animals in order to hunt them better.

In an earthquake-prone country like Japan, people must have noticed the odd behavior of animals before quakes a long long time ago. They must have noticed the behavior of catfish in particular. And from the restlessness of the little catfish, probably, came the idea of the restless, earthquake-causing big catfish.

## *Chapter Four*
# Tales of the Sky

Animal behavior has guided people in many ways. Noah, in the biblical story, used a raven and a dove to search for dry land after the Flood receded.

The Bible's story is very much like the tale of a ninth-century Viking navigator, Floki. He set sail from the Shetland Islands, north of Scotland, for Iceland, about 500 miles to the northwest. On board his ship he carried a cage of ravens.

A few days out from the Shetlands, Floki released a raven. The bird soared into the sky, circled a few times, and flew off in an easterly direction. Floki knew the bird was heading back for the Shetlands, though from his point of view they had long disappeared below the horizon. The bird's flight path gave Floki a bearing from his present position to the Shetlands.

Days later, he uncaged another bird. It soared, circled, and returned to the boat. No land was in sight.

More days passed. Floki knew his ship should be nearing Iceland. A third raven was sent aloft. This one headed west and did not return. Now the Viking sailor felt sure that Iceland lay over the western horizon.

Floki and Noah were taking advantage of the same two facts.

*Fact 1*: The higher you are, the farther you can see. To a man standing in a small boat, the ocean horizon is 3 miles away. But a bird 1,000 feet up can spot land nearly 40 miles distant.

*Fact 2*: Ravens and doves are land birds. Released over the ocean, they will try to find the nearest land.

Halfway around the world, in the Pacific, the Polynesians knew the ways of land birds and of sea birds, too. They did not take caged birds with them on their voyages from island to island. Instead, they used the natural habits of the birds as ocean signposts. Day by day, terns fly out from the islands to fish at dawn. They range as far as 20 miles to sea and return to their home islands as evening falls.

The birds, however, were not enough of a guide. The Polynesians made far longer voyages in their small boats than did the Vikings. They carried no maps of the islands that dotted their path. They had no instruments or charts to keep track of the stars wheeling endlessly overhead and no compasses to point their way.

How did they find their course across the great Pacific Ocean? The key is in old Polynesian stories. But the tales are not simple stories of birds and sailors, like the stories of Noah and Floki. They are tales whose meaning is more hidden.

One story tells of a Polynesian hero who sets out to travel from island to island in his long canoe. He sees an old woman at the door of her house and startles her. She ran away to the west. Next comes an old man in a canoe sailing toward the hero out of the west. The hero chats with him a while. Then he sails east toward the home of two old lepers.

The old woman, the man in the canoe and the two lepers do not walk the earth or sail the seas. Their paths are in the sky.

The old woman at the door of her house is a star group, in the constellation Taurus. We call this star group the Pleiades. It's a cluster of six naked-eye stars — seven, if your eyes are *very* good — in the shape of a tiny dipper. They're all packed into a space in the sky no bigger than the full moon, and a beautiful sight through binoculars, which show many more stars.

The old man in the canoe is the star Aldebaran, also in Taurus, near one tip of a V-shaped group of stars called the Hyades. The Polynesians saw this V-shape as a canoe. And the two lepers are the stars Castor and Pollux, the Heavenly Twins of the constellation Gemini.

Suppose for a moment that you are a Polynesian sailor cruising the

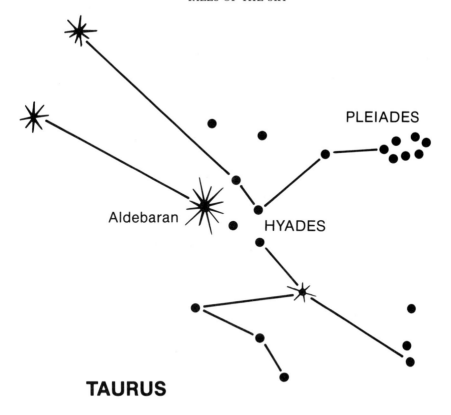

The Pleiades, an open cluster of stars in the constellation of Taurus the Bull. The Pleiades were believed by many people to influence the weather. Both the Pleiades and Taurus are part of the star maps passed down by word of mouth among the Polynesians. These verbal maps are still used by them for navigating.

Pacific just north of the equator. It is a clear, moonless, starry night in early November. Your course is a little east of northeast.

The Pleiades rise, and you steer your boat toward them. They rise, as they always do, east of northeast. Near the equator, the stars climb into the sky in a nearly vertical direction, so the Pleiades will lie straight above the point you're steering at for some time.

About an hour later, Aldebaran rises, marking your course. As the Pleiades "run away to the west," getting too high in the sky to be used as guide stars, Aldebaran takes their place. Nearly three hours later, the two lepers rise to take the place of Aldebaran and you steer by them.

Actually, Aldebaran rises a little to the south of the place where the Pleiades rose. Castor and Pollux rise a little to the north. But you know that and you make allowance for it. You know the stars in the sky as well as you know the islands in the sea.

The tale of the old woman, the old man, and the two lepers is only one of many Polynesian sky tales. They map the rising and setting places of many stars. These places form a circle around the horizon, a compass whose points are marked by stars.

Taken together, the stories provide a star map giving the positions of stars for any time of night and any season of the year. For centuries, Polynesian navigators have used the map to help them sail the world's largest ocean. It is a map drawn in story form and handed down from generation to generation by word of mouth.

The Polynesians still use the map unaffected by this age of radio and navigation by artificial satellites. So well do the stories chart the night sky that they can use the map even on partly cloudy nights. A break in the clouds and a glimpse of a few stars in a patch of clear sky are all these mariners need. From the position of these few stars, they know what stars are where on the horizon, even though the stars themselves are invisible.

Sometimes the navigators keep watch on an invisible island too — an island they call the "etak island." The etak island is a known island that lies out of sight many miles to the right or left of a boat's course. The Polynesians pretend that the etak can be seen. It appears to slide backward against the background of stars on the horizon, just as nearby scenery slips backward against the horizon when you are in a moving car.

So at one point in the journey, the sailor may say, "The etak is in line with the Pleiades." Later, as the boat passes due south of the etak, "The etak is in line with the North Star," and so on. In this way, he keeps track of his position with respect to an island that he knows. To help see this in his mind's eye, he imagines his canoe sitting

FIRST DAY

FIRST NIGHT

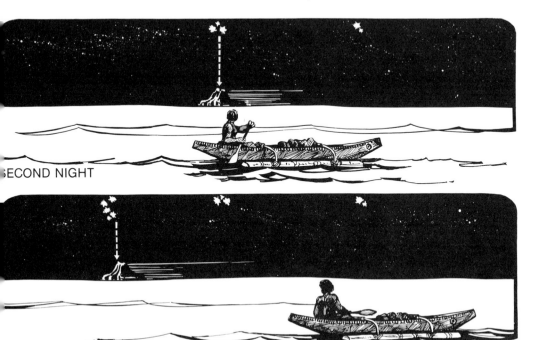

SECOND NIGHT

THIRD NIGHT

The etak is a known island that lies beyond the horizon of the Polynesian navigator. He pretends that his canoe stands still while the etak slides past him, lining up with one group of stars after another. (*Paul S. Plumer, Jr., Drawing*)

perfectly still in the water, while the etak slowly slides past him. At the same time, the Polynesians say, "Of course we know the island really doesn't move."

Scientists often do much the same thing. Modern star maps show the stars as if they were all pasted onto a bowl-shaped sky, the same distance away, and turning around the earth. In fact, some stars are much farther away than others, no stars turn around the earth, and stars move in many different directions. But they are so far away that their real motions are hardly noticeable, and they all appear to be at the same distance. For purposes of navigating on earth — or even within the solar system — it is useful to draw them as they appear on a star map.

Modern star maps also show the stars arranged into constellations. Long ago, these constellations were named for gods, goddesses, people, animals, and objects. Today, constellations are still a useful way of dividing the visible stars into convenient groups.

Our star pictures are not those of the Polynesians. They come down to us from the ancient Greeks and Romans. Like the Polynesians, they knew of the value of the stars for navigation, and they recorded their lore in tales.

The Greek poet Homer, hundreds of years before Christ, described how Odysseus kept his course for east and home: ". . . nor did sleep fall upon his eyelids, as he viewed the Pleiades and Boötes, that setteth late, and the Bear . . . which turneth ever in one place . . . and hath no part in the baths of Ocean."

Odysseus kept the Great Bear, which we know better today as the Big Dipper, on his left, thus keeping his bearing east. As for the Bear not taking a bath, Homer did not mean to imply that the beast was dirty; simply that it was so close to the North Star (Polaris) that it circled the sky without setting. To the Polynesians living north of the equator, the Bear does appear to set — to bathe in the ocean. To the Eskimos in the Arctic where the North Star and the Bear are both nearly overhead, the Bear circles round and round high in the sky. (That, by the way, is where the name "Arctic" comes from. *Arctos* is the Greek word for "bear," and the Arctic is the land where the Bear rides high.)

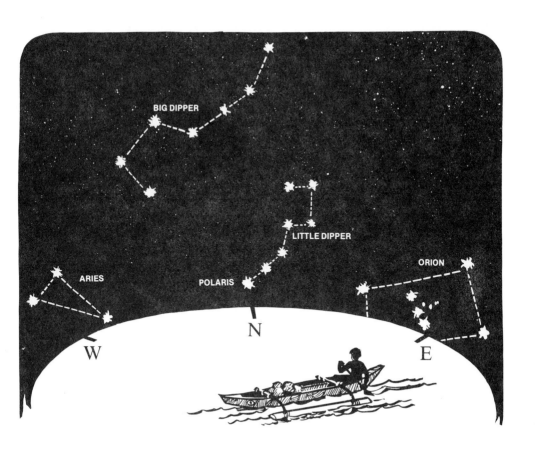

To the Polynesian navigator, the entire horizon is a compass — a compass whose points are lit by the rising and setting stars. For example, the two stars at the end of the Big Dipper's bowl always point toward the North Star (Polaris) and the north point of the horizon. Aries will always set a little north of west and rise a little north of east. The three stars making up Orion's belt will rise nearly due east and set nearly due west. Aries does not actually set as Orion rises. But at any time or season, the Polynesian navigator can use some group of rising or setting stars to mark a direction on the horizon. *(Paul S. Plumer, Jr., Drawing)*

Another star group important to the ancient Greeks for navigation was the Pleiades. When the Pleiades rise in the east just before the sun, it is May. To the Greeks, this predawn rising of the Pleiades was a sign of good sailing weather in the Mediterranean Sea.

Six months later, in November, the Pleiades rise in the east as the sun is setting. They are high in the sky at midnight and visible all night long. November is a time of wind, storm, and rain in the

Mediterranean. To the Greeks, the Pleiades high in the sky at midnight meant that the time of bad sailing weather had begun.

The fact that the visible constellations change regularly with the seasons gave ancient peoples a kind of natural calendar in the sky. Today, the stars are hidden for most of us by city lights and smog. Calendars are pads or sheets of papers with numbers and days neatly stacked. Time is the tick or hum of a watch, the jittery flight of numerals on the face of a digital clock.

But overhead the stars turn and change with the seasons as they always have. In observatories around the world, time is still measured by the moment that a given star passes across a certain imaginary line in the heavens, the meridian. And in ways we do not often think of, we still feel the influence of an older view of time, stars, and seasons — a view that was part fact, part fancy; part science, part myth.

Most of our star lore is handed down from peoples who lived north of the equator. For them, the rising of the Pleiades just before sunrise meant springtime and good growing weather. When the Pleiades were high at midnight and in the sky all night, it meant a time when life was at low ebb, getting ready for the long sleep of winter.

Growing and going to seed, birth and death, the beginning of things and the end of things — all were linked to the Pleiades. November was the month of the Pleiades. For many peoples, it was the time when the old year ended and the new year began, when great tribal meetings were held and important decisions made. Because life was at low ebb, it was also a time for remembering the dead.

To this day, we still celebrate Hallowe'en on November's doorstep. The long tradition of November meetings was one reason the Founding Fathers fixed the United States elections for November.

South of the equator, too, the Pleiades are important calendar stars. But there the seasons are reversed. In Australia, most of South America, and Southern Africa, it is winter when we have summer, spring when we have fall; November, the Pleiades month, is in the warm time of the year.

For this reason, many peoples in the Sourthern Hemisphere once believed that the Pleiades caused hot weather. When they were in the sky most of the night, it was hot. The sun was in the sky every day, summer and winter, so it could not be the cause of hot weather.

The argument makes a certain amount of sense. It does recognize that there is a connection between the position of the Pleiades in the sky and the coming of hot weather.

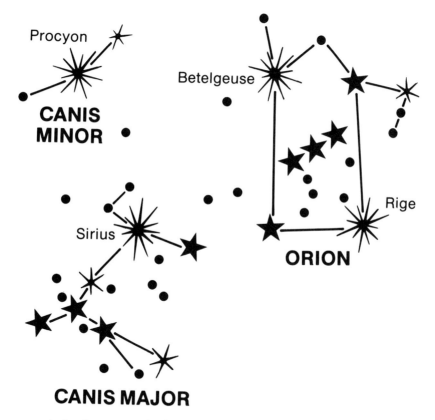

Sirius, the brightest star in the sky lies in the constellation of Canis Major, the Great Dog. The star is in a line with the three stars in the belt of Orion the Hunter. Bright Sirius, rising with the sun in summer, was once thought to be the reason for summer heat.

Peoples in the Northern Hemisphere also thought that certain stars were the cause of summer. About 2,400 years ago, the Egyptians noticed that when Sirius rose in the east just before dawn, it marked the start of the hottest time of the year. At that time, Sirius and the sun would both be shining down on the earth during most of the day. (Of course, Sirius would not be visible in the day sky because of the brightness of the sun.)

The Egyptians thought the heat at that time of year was due to Sirius adding its rays to the sun's rays. Sirius is the brightest star in the night sky, and its brightness, added to the glare of the sun, tipped the scales toward summer.

Or so the Egyptians said. The Romans had the idea. Because Sirius is in the constellation of Canis Major, the Great Dog, the Romans used the phrase *dies caniculariae*, the dog days, for the hottest part of the year, an expression we still use.

The Greeks believed much the same. Almost four hundred years before Christ, Hippocrates, some of whose writings on medicine were centuries ahead of their time, wrote of Sirius' power over the weather. And during hot weather, he pointed out, sickness and epidemics were more widespread. Therefore, it was argued, Sirius had great influence over health.

These beliefs, like those of Southern Hemisphere peoples about the Pleiades, had things backward. Another ancient Greek, Geminos, who lived around 70 B.C., put his finger on the flaw in the idea when he wrote:

"It is generally believed that Sirius produces the heat of the dog days; but this is an error, for the star merely marks a season of the year when the sun's heat is greatest."

Geminos was an astronomer. However, the beliefs of most ancient peoples about the stars and other heavenly bodies came a lot closer to astrology. The stars, sun, moon, and planets control people and events here on earth, including the weather. The study of weather was a mere stepchild of astrology.

In fact, up to the seventeenth century there was no notion that our planet has an atmosphere. The "air" around us with its clouds and winds was no different from the empty space in which the heavenly bodies moved. The great Greek philosopher Aristotle's classic work *Meteorologica* is about things high in the air — the motions of the stars and the causes of the weather.

That's where we got our modern scientific term for the study of the weather, "meteorology." It has nothing to do with meteors. "Meteor" simply comes from the same Greek word as "meteorology" does, *meteoron*, meaning "something high in the air."

Nowadays we know that the earth does have an atmosphere and it is in the atmosphere that our weather changes take place. But though sun and moon lie far beyond the atmosphere they can tell us about changes in it. In this way they can be weather forecasters.

The Zuñi Indians of the American Southwest say, "When the sun sets unhappy, the morning will be angry with storm." This is a weather rule that will be true most of the time.

The "unhappy" sun is a sun dimmed by thickening haze and clouds. If this happens near sunset, the cloud cover is west of you. (Around the world in the temperate zones, storms, like the weather, usually move west to east.) Slowly thickening clouds often mark the edge of approaching storms, especially big ones that bring heavy rain or snow. So an "unhappy" sunset tonight may well mean "unhappy," or "angry," weather tomorrow morning.

Similar ideas are expressed in weather lore all around the world and are very old. The biblical version is: "When it is evening, ye say, It will be fair weather: for the sky is red. And in the morning, It will be foul weather to day: for the sky is red and lowring."

Another version is: "Evening red and morning gray/Help the traveler on his way/Evening gray and morning red/Bring down rain upon his head."

When the sun shines through dry air, fine dust particles scatter the blue part of the sun's rays around the sky, giving it its blue color. Since the blue rays are filtered out, the sun itself appears yellowish.

Near the horizon, in early morning and late afternoon, the sun must shine through more air than when it is overhead. If the air is dry, the blue, yellow, and green rays are filtered out by the dust.

The sun then appears red. If the air is full of water droplets from clouds or fog, the sun appears grayish.

"Evening gray and morning red" means that rainy weather lies to the west, toward sunset, and is approaching; fair weather lies to the east, toward sunrise and is going away. Fair weather is leaving, bad weather is moving in: that is that message of sun, cloud, and sky. "Evening red and morning gray" means just the reverse.

Generations of weather experience lie behind these sayings. They work well in the world's temperate zones where storms do generally travel from west to east. They don't work in the tropics, however. There, storms follow the local winds which may come from any direction. So sailors, traveling from temperate zones to the tropics and back again, made up another version: "Rainbow to windward, foul fall the day/Rainbow to leeward, damp runs away."

This version of the rule adds some wisdom of its own: For a rainbow to form, sunshine and rain must both be within sight. The bow usually forms at the edge of a storm area. "Rainbow to windward" means the storm is in the same direction as the wind and moving toward you. "Rainbow to leeward" means the storm is in the opposite direction from the wind and moving away from you.

This works equally well in other parts of the world. With new knowledge thrown in, the sailors' version sharpens the old rule, much as new evidence can sharpen an older theory in science.

Like the sun, the moon gives advance warning of weather changes: "If the moon show a silver shield/Be not afraid to reap your field/But if she rises haloed round/Soon we'll tread on deluged ground."

"Silver shield" simply means the moon is shining in a cloudless sky. "Haloed round" means a ring of light around the moon. Such a ring appears when a high, thin veil of cloud moves slowly over the sky.

High thin cloud like this, slowly thickening, is usually a sign of a large storm approaching. Often, the first veil may be far ahead of the storm, which will eventually bring prolonged rain or snow. This storm is the result of warm air moving in and riding up over a mass of heavier, colder air.

On the other hand, a sudden shower without much cloud warning is usually the result of a fast-moving "bundle" of cold air and is soon over. This was noticed by people long before they knew anything about warm and cold air masses: "Rain long foretold, long last/Short notice, soon past."

These rules of weather change are general rules, true over most of the world. But much weather lore is very local and goes into great detail about conditions in one place.

The Finns, for example, have many words for ice, one for each type.

There is the first thin, elastic film of ice when freezing begins in fall. Then there is the "ice rind" when the film reaches an inch to 1½ inches in thickness. Later, the ice rind thickens into young winter ice, tough and strong but translucent: light can pass through it as if it were a piece of frosted glass. And finally there is deep winter ice so thick that no light can pass through it. All of these forms of ice and others have different names in Finnish.

In the same way, the Eskimos in the North American arctic have many names for snow. If it is the blinding snow driven along by a gale, it has one name. If it is loose-packed snow on the ground, it has another. If it is packed together in hard solid masses — a sort of cement made of snowflake crystals glued together by ice, good material for building igloos — it has a third name.

This sort of careful classifying is an important part of science, too. Sometimes scientists have borrowed words from one language or another to help in classifying. Geologists, for example, borrow from Hawaiians. The Hawaiians, who have had long experience with volcanoes, have two names for lava: *aa* (AH-ah) and *pahoehoe* (pa-HO-ee-HO-ee). *Aa* is lava in a jumbled field of sharp, jagged pieces of volcanic ash, often almost impossible to walk over. *Pahoehoe* is hardened lava with a smooth wavy surface like a frozen ocean. Now both *pahoehoe* and *aa* have become scientific terms.

As we have seen, the Finns and the Eskimos have many names for ice and snow. From another point of view, of course, ice and snow are both just water. So are clouds. They are suspensions in the atmosphere of countless numbers of tiny droplets held aloft by air

Hawaiians gave the name *aa* to lava beds like these. *Aa* is now a word of science, used by geologists around the world. (*United States Department of the Interior, National Park Service Photo*)

In the valley below Haleakala, on the Hawaiian island of Maui, water fills low-lying spaces in old lava beds, forming a chain of pools. This lava has hardened into a smooth surface called *pahoehoe*. *Pahoehoe* is another Hawaiian word that has become a word of science. (*United States Department of the Interior, National Park Service Photo*)

currents. You can see that on a plane trip when you fly through a cloud; you can also often feel the bumpiness of the currents. Or you can feel the mist on your face when a cloud forms at ground level and becomes fog.

Just as the Eskimos have many words for snow, so people have seen many kinds of clouds. We've already talked about the high, thin veil of clouds that forms a halo about the moon and foretells a storm. Scientists now call these "cirrostratus clouds."

One cloud saying describes a different type of cloud: "If woolen fleeces spread the heavenly way/Be sure no rain disturbs the summer day." That's a perfect description of the typical fair-weather cloud of summer — the "cumulus cloud."

Sometimes, however, these clouds build up into yet another type of cloud which is very threatening. Here's how a centuries-old bit of weather lore puts it: "In summer . . . when it grows very hot, and you see clouds rise with great white tops like towers, as if one were upon the top of another, and joined together with black on the nether side, there will be thunder and rain suddenly."

This is an accurate picture of fair-weather clouds on a hot sultry day growing into towering thunderheads — "cumulonimbus clouds."

Folklore has described cloud types for centuries. Not until around 1800 did scientists begin naming them. A French scientist, Jean Baptiste Lamarck, was trying to set up a system of weather forecasting. At this time, he had become too poor to afford many instruments, so he had to depend largely on what he could see with his own eyes. He watched the clouds float by his garret window and noticed that different cloud patterns led to different kinds of weather. He saw the need and the use for sorting clouds into weather types. Today, the charts used by airplane pilots and meteorologists list twenty-seven types of clouds.

There are many possible ways of describing and sorting clouds or stars or anything else of interest. What ways we choose depend partly on what we think is useful at the time.

With new needs, a new viewpoint must be found. With a new viewpoint comes new knowledge. The Finns, the Eskimos, and the

All the basic cloud types are in this sky. Near the top are wispy streaks of cirrus. The white cloud with the flattened, anvil-shaped top is a developing thunderhead — a cumulonimbus. Below it, long layers of low-lying stratus stretch across the photo. At the bottom are flocks of puffy cumulus clouds. (*National Center for Atmospheric Research Photo*)

Cumulonimbus clouds. This thunderstorm was photographed on a hot June day north of Denver, Colorado. (*National Center for Atmospheric Research Photo*)

Hawaiians have shown us that ice is more than "ice," snow more than "snow," and lava more than "lava." Cloud lore and cloud science reveal that when you've seen one cloud you haven't seen them all. As old tales about animals and lights in the sky have given scientists messages about earthquakes, old tales about clouds have given them messages about weather.

*Chapter Five*

# Medicine Has Roots

## I. The Stone Age Flower Power

There is a cave in Iraq called Shanidar, high in the northern tip of the Zagros Mountains. The Neanderthal people buried some of their dead in Shanidar 60,000 years ago. They buried one with flowers.

The Neanderthals lived in Europe and parts of Asia from about 75,000 to 35,000 years ago. They were a tough people, who managed to survive through the bitter cold of the next-to-last Ice Age. They made good stone tools, used fire, and hunted herds of reindeer over the snowy meadows.

Yet in the end, the Neanderthals disappeared. About five thousand years after Cro-Magnon man came on the scene, the Neanderthals died out. The Cro-Magnons had better tools and weapons. They lived, worked, and hunted in large groups held together by special rules and ceremonies. When the two peoples lived in the same place, the Cro-Magnons got most of the food and may have driven the Neanderthals away from the best hunting grounds. But there is evidence that some Neanderthalers picked up the Cro-Magnon arts of tool-making and learned them very quickly. Possibly Neanderthal and Cro-Magnon lived together occasionally. If so, those of European stock may have a few Neanderthal ancestors, since some anthropologists claim modern Europeans to have descended from the Cro-Magnons.

Once upon a time that idea would have been insulting to most

Caucasians. The first Neanderthal remains were discovered in the Neanderthal Valley, now in West Germany, in 1856. Over the next fifty years, many more similar remains turned up in Europe, together with stone tools and the bones of woolly mammoths and other extinct animals.

Neanderthals' bones showed that, by modern standards, they had rather low foreheads, bulging cheekbones, and receding chins. They were short and thickset, with massive arms and legs. The size of their skulls, however, indicates that they had modern-sized brains.

The world the Neanderthals lived in may explain much about their appearance. The bulging cheekbones supported the powerful jaw muscles needed for a diet of tough meat. Most people who live in very cold places are stockily built. A short, compact frame conserves body heat better.

Nevertheless, many nineteenth-century scientists regarded the Neanderthals as being subhuman — more ape than man. That picture began to change with the finding of the first formal burial sites in the early 1900s. At one site, a teen-age boy had been put into a shallow grave lying on his right side, with his head on his arm as if he were asleep. Near his hand was a stone ax. Around him lay animal bones burned in a fire — food, perhaps, for a long journey in the afterlife.

Neanderthal man may not have looked as "modern" as Cro-Magnon man. But what he felt about life and death was undeniably human. And each new discovery of burial places has made him seem even more so.

The flowers at Shanidar have long since withered into dust. But they have left behind traces as lasting as bone, as clear as a written record. Clusters of fossilized pollen lie scattered thickly around the remains of the man buried there. Each kind of pollen is the "fingerprint" of a specific flower.

Eight different species of flowers were buried with the Neanderthal man at Shanidar. Among them were large masses of hollyhock. Unlike some of the other flowers in the grave, hollyhock does not grow in bunches. Before the man's burial, someone had to pick the hollyhocks one by one. Dr. Ralph S. Solecki, the American anthropologist who discovered the grave site in 1953, thinks this is strong evidence that the flowers were gathered purposely for the burial. Probably the dead man had been important, a leader or a medicine man. Maybe the plants were a part of his stock of medicines.

In fact, the Iraqi Ministry of Agriculture now lists hollyhock as a valuable medicinal plant. Solecki notes that in Iraq the roots, leaves, flowers, and seeds are used to make a variety of medicines for the relief of toothache, inflammation, and dressing wounds. "It seems to be the poor man's aspirin," Solecki writes. And six of the remaining seven plants are also on the Iraqi list of medically useful plants!

Writing as a careful scientist, Solecki remarks: "Naturally we cannot be sure that whoever buried [the man] was aware of the . . . medicinal properties of the flowers . . . But it is extremely likely that, as practicing naturalists (and early-day ecologists?), the Neanderthals must have known and appreciated all of their environment, since their very existence depended on it."

Of course their experience with useful and poisonous plants would have been passed down the centuries by word of mouth, just as the star lore of the Polynesians was. But the Neanderthals are extinct, and only the pollen in scattered graves can speak for their plant lore. The rest is lost.

Primitive peoples around the world in all ages have had broad knowledge of the medical and food values of plants. Slowly, that knowledge has become a part of folklore. It reflects some of the human spirit of science — the search for order and truth and understanding. The folklore may be partly superstition, but there is most often a kernel of hard fact in it.

We stand in danger of losing some of that knowledge as we have lost the lore of the Neanderthals. Botanist Dr. Siri von Reis Altschul points out that there are several thousand plants valued as medicinal in current folklore of many peoples. Less than half of these plants have ever been studied scientifically. Some of the rest may become extinct before science gets to them — as may the folklore and the peoples themselves. She suggests that the Western world's great herbariums be explored for useful medicinal and food plants before it is too late.

In herbariums, plant specimens are stored along with notes made by the scientists who collected them. These notes often include references to folklore and local beliefs about the value of the plant. Altschul and her colleagues spent four and a half years searching through the 2,500,000 plant specimens of the herbarium at Harvard University and found over 5,000 potentially useful plant species.

## II. The Poison Cure

In a way, Altschul is only suggesting that science take the path that medicine itself has followed. Many plant remedies that began as folklore are now part of medical science. An example is one of the oldest plant-derived medicines known — colchicum.

Colchicum is made from the seeds and roots of the autumn crocus or meadow saffron, an herb of the genus *Colchicum*. It is a deadly poison — and a uniquely useful medicine.

Colchicum's fame as a source of poison goes back beyond history into legend. The very name of the drug comes from Colchis, an ancient Asian country, home of the sorceress Medea and the land of the Gold Fleece of Greek mythology. Medea fell in love with Jason, the Greek hero. She helped him get the Golden Fleece, betrayed her father, killed her brother, and fled with Jason to Greece.

Years later, Jason deserted Medea and their two sons, planning to wed the daughter of King Creon of Corinth. Medea took a horrible revenge. She soaked a beautiful robe in colchicum and sent it to Jason's bride as a gift. When the princess put it on, she died in agony as if she were being burned alive.

In fact, colchicum must be taken internally to do its lethal work. But the symptoms, including burning sensations in the throat and stomach, are every bit as terrible as the legend makes them out to be.

Yet colchicum had another side. The Egyptians wrote of it as a source of medicine in the seventeenth century B.C. In Europe, where the autumn crocus is a common wildflower of the meadows, colchicum has long been used for gout, an excruciatingly painful inflammation of a joint. Usually, the first attack of gout occurs in the big toe, but it may also hit the ankle, wrist, knee, or elbow.

Gout is caused by uric acid crystals forming in the joint. Normally, these crystals remain dissolved in the body fluids. But in some people — partly because of their diet and partly for other reasons — the crystals are deposited in the joint. The deposits cause pain, inflammation, and swelling. Eventually, the bone and tissue of the joint may be destroyed.

Colchicum, taken in very small doses, relieves the pain of gout. It does not *cure* the disease. We now have other medicines that can control gout. But to this day, colchicum is the only medicine known that relieves the pain of gout quickly and effectively.

The first known physician to use colchicum in treating patients with gout was Alexander of Tralles. He lived in the Eastern Roman Empire during the sixth century A.D. Nevertheless, it was not until the nineteenth century that most doctors accepted colchicum's medicinal properties.

There were many reasons why it took so long. Colchicum was a folk remedy, and folk remedies were supposedly mere superstitions. It was peddled by wandering herb "doctors" and quacks. Moslem physicians, who were strong on herbal medicine, rediscovered it in the tenth century. But European doctors would have none of it. In their eyes, Moslem medicine was not to be trusted.

Worst of all, colchicum was a dangerous poison. It was hard to extract from the roots of the crocus, and the strength of the extract varied from plant to plant. So the proper dose for one batch of colchicum might turn out to be a lethal dose of the next batch.

In 1884 a French chemist named Houdé succeeded in making the pure drug, colchicine, from colchicum extract. From then on, doc-

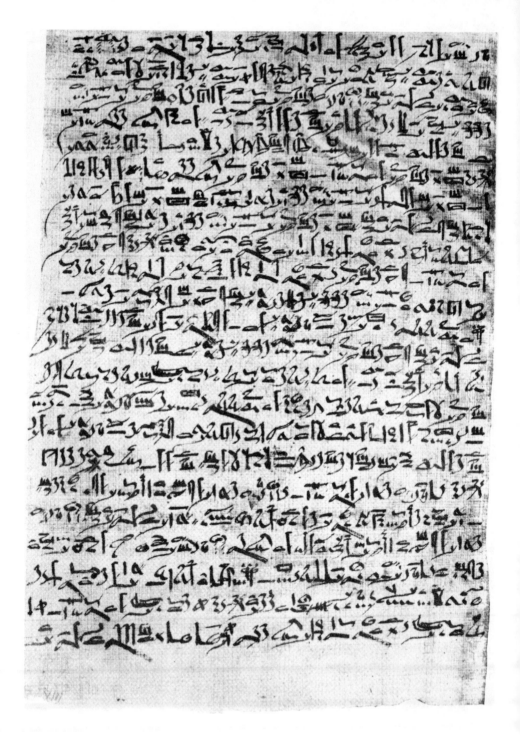

A page from an Egyptian medical book of the seventeenth century B.C. (*New York Public Library Picture Collection Photo*)

tors could give each gout patient an exact dose of the pain-relieving medication. Today, many plant-derived medicines are purified in the same way. Others, originally derived from plants, are made artificially in the laboratory. More than half the prescriptions filled in the United States today contain plant-derived drugs or their laboratory substitutes.

## III. Horsetails, Hay Fever, and Headaches

That's true of many non-prescription medicines too. Have you ever used an inhaler or nose drops to get relief from a stuffed-up nose? Then you were probably using another very ancient plant drug, ephedrine.

For more than 5,000 years, the Chinese herb *ma huang* has been a part of Chinese medicine. It was used for treating asthma and some other problems of the respiratory system. Only within the past hundred years has Western medicine discovered that the drug derived from *ma huang* — ephedrine — is in fact useful for such ailments. Asthma, for example, may sometimes be due to an allergic reaction such as hay fever. Ephedrine is commonly used to relieve hay fever nowadays.

It's an interesting sidelight that *ma huang* is not only the source of one of the oldest known plant medicines but that the plant itself is practically a living fossil. Plants like *ma huang* were common around the world 280 million years ago, when the early ancestors of the dinosaurs were just getting their start. *Ma huang*'s living relatives in this country are the joint fir and the Mormon tea bush of the American desert areas.

Notice that name, "tea bush." Ephedrine, like the caffeine in coffee and tea, is a stimulant. The Mormon tea bush, also called "Brigham tea" after the Mormon leader Brigham Young, was used as a tea substitute.

And it's the appearance of the plants that gave the drug ephedrine

Medieval doctors relied heavily on medical lore from ancient times. They used a mixture of magic, common sense, shrewd observation, and guesswork in their work. (*New York Public Library Picture Collection Photo*)

its name. They look something like an animal tail with many branches that end in tufts. The Greek name of the genus, *Ephedra*, simply means "horsetail."

Aspirin, the most common medicine of all, gets its name from a genus of plants in a somewhat roundabout way. The story of aspirin winds back and forth between folk medicine and medical science. It is a trail blazed by accident, guesswork, and an occasional stroke of genius.

American settlers learned much herb lore from the Indians on their long journey westwards. One popular remedy was a drink made from the bark of willow trees. It relieved fever and headaches and was good for aching joints. Around the world, willow bark has been used as a medicine for thousands of years.

Medical science paid little attention to this folk remedy, however. Here and there doctors heard tales from their patients of some old family recipe for aches and pains that included willow bark. They heard stories of how well it worked, but they remained skeptical. Folk tales are not science and there was no hard evidence that willow bark had any medical uses. Moreover, people will believe what they want to believe. If they believe a remedy is going to make them feel better, they often *will* feel better even if the remedy is worthless. Or they will *believe* they feel better. People may believe what they see, but as often as not, they see what they believe.

Nevertheless, nineteenth-century chemists *had* found a use for willow bark. It contained a white, sweet-tasting powder useful for making dyes. Despite its sweet taste, the powder is an acid. The chemists named it "salicylic acid," from *Salix*, the Latin genus name for "willows."

Salicylic acid made from willow bark was expensive. In the 1850s, a German chemist, Dr. Hermann Kolbe, discovered how to make salicylic acid out of carbolic acid in the laboratory. This was a much cheaper method.

Kolbe also noticed something else. Salicylic acid slowly turns back into carbolic acid again in the test tube. That was interesting, but not particularly useful. Kolbe filed the fact away in his memory.

Carbolic acid, or phenol, as it's technically called, has definite uses in medicine. While Kolbe was making his discoveries, the English surgeon Joseph Lister, was proving that infections in operating rooms often spread from one patient to another because of poor hygiene. Rooms, clothing, and instruments, he said, had to be kept clean and free from infection-causing germs to prevent the spread of infection. And phenol, Lister demonstrated, was an excellent germ-killer.

Unfortunately, you can't just swallow a dose of phenol and get rid of any bug that's bugging you. The acid would kill the germs all right. But it would also kill you; it's too strong to take internally.

Was there a way to make phenol safe to take? A German surgeon, Karl Thiersch, wondered about that. Perhaps, he thought, some substance could be found that would turn into phenol in the body very slowly — too slowly to harm the body, but fast enough to destroy germs.

Thiersch was a friend of the chemist Kolbe. When he talked to Kolbe about his problem, Kolbe remembered his own discovery. That must be the answer! After all, Kolbe said, the body is like a big chemical factory. It's nothing but a test tube. In a wild rush of enthusiasm, Kolbe set out to prove this was true.

He and his colleagues tried salicylic acid on milk. It stayed fresh for a week at room temperature. Salicylic acid kept meat from decaying and wine from going sour.

That showed, Kolbe reasoned, that the salicylic acid slowly turned to phenol in the milk, meat, and wine. And the phenol killed the germs.

The news spread. All of a sudden, salicylic acid was the wonder drug of medicine. Doctors gave it to patients with typhoid fever. The patients felt much better. Doctors gave it to patients with pneumonia. They felt better. Doctors gave it to patients with rheumatism. They felt better. And so on.

There was one minor flaw: most of the patients with grave diseases like typhoid or pneumonia died, even though salicylic acid made them feel better. The doctors on the salicylic-acid bandwagon had an explanation. These patients didn't get enough of the medicine. Or they didn't get it in time.

Doctors, like the rest of humanity, may occasionally believe what they want to believe.

One doctor, at least, did not believe. He was Carl Emil Buss of Switzerland. If salicylic acid killed germs, why did the patients with dangerous diseases like typhoid and pneumonia die? If salicylic acid had no effect, why did they feel better?

The answer to the second question was easy. They felt better because their fevers had gone down. But what did salicylic acid have to do with that? Quinine was the great fever-reducing drug. Of course there were folk tales about things like willow bark being used to cut down fever. But . . .

Wait a minute. Where did salicylic acid come from anyway? There was something about the name of the willow genus, the scientific name . . .

Buss looked it up. The same was *Salix*. Now he remembered. Salicylic acid came from salix plants — from willows. It was beginning to look as if the folktales had something.

They did. Buss tested his idea slowly and carefully with many many patients. The willow-bark drug had found its way from folklore to medicine cabinet. But its journey was not yet over.

Lots of people didn't like to take salicylic acid. It burned their throats and gave them upset stomachs. One of them was a German named Hoffman who took the medicine for his aching joints. He wanted his son, Felix, to find a better painkiller. So did Felix's bosses at Bayer & Company, the great German chemical firm.

So Felix began trying to change salicylic acid into another substance — a painkiller without the bad side effects. After some time, in 1893, he found a compound he called acetylsalicylic acid. And it fit the bill perfectly.

As drug manufacturers do today, Bayer looked around for a good short name for its new product. "Acetylsalicylic acid" was too long.

How about "a-salicylic"? No, that was still pretty long and complicated. Besides, it would remind people of salicylic acid.

Then Felix Hoffman remembered something from herb lore. There was anther group of plants that were sources of salicylic acid, plants like the flowering shrub meadowsweet. The scientific name of their genus was *Spiraea*. That put an idea in his mind. Why not call the new drug "a-spirin"?

## IV. The Idea of a Germ

Today we know that salicylic acid slows down the growth of germs in foods. That's why Kolbe found that it keeps milk, meat, and wine from spoiling. But salicylic acid does *not* turn into phenol in the foods. Nor does it do so in the body. Reactions in the body are far more subtle and complex than reactions in a test tube.

Nor does salicylic acid *kill* germs. We've seen what happened when doctors dosed dangerously ill patients with it. In addition, some doctors tried dressing infected wounds with salicylic acid. They may have gotten the idea from Kolbe's experiments. But the results were generally disastrous.

Dressing wounds to help them heal is a very old practice. Egyptians used a dressing made of moldy bread as long ago as 1900 B.C. So did the Chinese and the Indians of North and South America.

Molds are plants that must grow on other living things or on their remains. They were used for dressing wounds because the idea worked. Long experience showed that wounds treated in this way were less likely to become infected. But no one could say why it worked.

By the end of the Middle Ages, most doctors were ignoring the idea of using molds to dress wounds. It seemed just one of the many superstitions so common at that time. And molds could have nothing to do with the way disease and infection spread.

Hippocrates, the great physician of ancient Greece, had long ago explained the causes of disease and epidemics. In those times, Greece suffered terribly from epidemics of malaria. Hippocrates himself may have been a victim of the disease.

He pointed out that the epidemics came at the hot time of the year. When the rains filled the marshes and the wind blew from the marshes, malaria came. Bad air, which the Greeks called *miasma*, caused the fever.

A medieval view of the four humors (earth, air, fire, and water) and how they affect the health of the body and the mind. Everyday language still reflects this idea that the "humor" of a person affects his or her behavior. (*New York Public Library Picture Collection Photo*)

The bad air entered the body and separated the hot and cold elements — the four "humors," as Hippocrates called them. So long as the humors were well mixed, the body was neither too hot nor too cold. When the humors were separated, illness resulted.

Hippocrates' theories of the cause of disease and epidemics influenced doctors' thinking for more than 2000 years. In fact, the very name "malaria," coined by a doctor in the 1600s, comes from the Italian for "bad air."

Yet there were other ideas. In ancient India, a malaria-ridden country, a medical writer and poet living in A.D. 300 wrote:

> The green and stagnant waters
> > like his feet
> And from their filmy iridescent
> > scum
> Clouds of mosquitoes, gauzy
> > in the heat
> Rise with his gifts:
> > Death and delirium.

Girolamo Fracostoro was a sixteenth-century Italian doctor and, coincidentally, also a poet. He had strong doubts about the theories of Hippocrates. He noticed that many diseases seem to spread by bodily contact. Seeds of disease, said Fracostoro, grew in the ill person, then spread to others.

What seeds? Who had ever seen them?

By the next century, someone had: the first microscopes were invented. A German Jesuit scholar, Athanasius Kircher, when a professor at the University of Würzburg, acquired one of the best of them. Kircher was a daring man. The deadly black plague was raging in Europe. The professor took a drop of fluid from an open sore on a plague victim. Under the microscope he saw "tiny worms" obviously alive and swarming in teeming millions. "Living contagion," he called them.

But there was no way in Kircher's time to do the careful experiments needed to show that germs can cause disease. Only in the late nineteenth century did that become possible. Then the work of men like Louis Pasteur of France, Joseph Lister of England, and Robert Koch of Germany showed that bacteria and other microorganisms were indeed the cause of many kinds of infectious diseases. Koch made the classic experiments to prove that one germ was the cause of each disease — one for cholera, one for tuberculosis, and one for anthrax.

He laid down the clear and simple rules for proving this connection for any disease: (1) show that the germ was present in ill people and not in healthy ones; (2) grow the suspect germ in pure cultures in the laboratory, unmixed with other germs; and (3) show that healthy animals injected with that kind of germ alone got the disease.

At last, Western science had the kind of precise answers it likes: one germ, one disease. And things seemed very clear.

The germ theory won out over the old ideas put forward by Hippocrates. The result was a powerful tool for the prevention and even the cure of many deadly sicknesses. It was a new way of looking at the problem of disease, a way far closer to what really happened when a healthy person caught an illness from a sick one than Hippocrates' view had been.

But the new way was not the final truth. And Hippocrates still deserves the title "father of modern medicine," though some of his ideas proved mistaken. For Hippocrates took the magic out of medicine and made it a science. He was the first to write that disease was not caused by evil spirits, nor was it a punishment from the gods. Look at the patient. See what's going on and try to understand it. Then decide on treatment. That's what the physicians who followed Hippocrates taught.

Many of his observations were shrewd and accurate. Malaria *did* spread in the hot months. It *did* follow the winds off the marshes. Hippocrates even noticed that there were several types of malaria, depending on whether the fever and chills occurred every second or every third day. He missed the true causes — a one-celled animal parasite spread by the bite of certain mosquitoes. The parasite could not be seen by the naked eye. Even had he guessed at the mosquito connection, as did the Hindu doctor-poet, he could not have proved it. It was beyond the limits of what he knew and what he could know.

In spite of the guesses of Fracostoro and the eyesight evidence of Kircher, the proof of the germ theory was still beyond the knowledge of doctors in the sixteenth and seventeenth centuries. *Within the limits of what they knew*, the Hippocratic theories of disease and epidemics made sense. Fracostoro's daring vision of "seeds of disease" did not.

What's more, the doctors of that time were well out from under the long reign of superstition that had dominated the Middle Ages. They were wary of such notions as the curative powers of powdered unicorn's horn. (That was just as well, since there are no unicorns.)

They scoffed at the notion of carrying bulbs of the autumn crocus in one's pocket as a good luck piece. (But they missed the hazy connection between this mistaken bit of folklore and the fact that the autumn crocus bulbs *did* contain a valuable medicine.)

They could not believe that moldy bread was of any value in preventing wounds from becoming infected. (And so they threw out a useful piece of folk medicine along with the nonsense and half-truths.)

Suspicion of folk medicine became a kind of superstition in itself.

Nobody knew why folk medicines worked. They had an air of magic and mystery about them. People who practiced folk medicine were suspected of being sorcerers who dealt in black magic.

Most of these people were women, since women were primarily concerned with family care. They were among the hundreds of thousands of women reportedly put to death as witches from about 1480 to 1780 in Europe alone. More than useful folklore was lost through this superstition.

## V.  Moldy Medicine!

Life is complicated. Science tries, when possible, to be simple. Sometimes it is not possible. Too much useful information is lost. Or the scientist looks too narrowly at his problem, seeing only what he expected to see.

For example, with the triumph of the germ theory, the way was clear for new ideas about preventing infection. One was Lister's idea of using phenol as an antiseptic to destroy infection-causing germs. Another might have been the idea that some molds produce a substance that kills certain harmful germs.

Lister himself might have discovered this in 1871. He tried to grow bacteria on dishes containing nutrient broths. One of his test dishes became covered with mold.

Would germs grow in the midst of this mold? Lister put a very small amount of the mold in another dish, leaving most of the mold where it was. He added disease-causing bacteria to each of the two dishes. In the dish with the tiny amount of mold, the bacteria ran rampant, moving about vigorously. In the other dish, bacteria soon stopped growing and spreading. They were alive, but motionless, almost as if paralyzed. But at that point, Lister dropped the whole experiment.

Five years later, a British physician, Dr. John Tyndall, tried to grow bacteria and a mold in one test tube. The mold formed a dense plug at the top of the tube and most of the bacteria died.

Why? Tyndall decided it was because the plug of mold had cut off the supply of oxygen to the bacteria and let it go at that.

For the next forty-six years scientists kept making accidental discoveries about the interaction of molds and germs and they kept missing the point. The scientists ranged from established professionals to students in the laboratory.

Sometimes the findings were shrugged off or ignored as off-beat results of no significance. (Off-beat results do happen in science. A colleague of Robert Koch's swallowed a whole beakerful of cholera germs to disprove the germ theory. He suffered no ill effects. Others who imitated him died in agony.) And sometimes students were scolded by their professors for allowing their culture dishes to become contaminated with molds.

The same sort of accident happened to Dr. Alexander Fleming in St. Mary's Hospital Medical School, in London. During World War I, Fleming had worked hard trying to heal infected war wounds, and after the war he continued his work. In the fall of 1928 he was growing staphylococcus germs in culture dishes. Staphylococci cause carbuncles, boils, and sometimes fatal infections of wounds.

One dish had become contaminated with mold. Clumps of bacteria were growing on the mold-free part of the dish. But between the mold and the germ colonies there was a clear area of broth, like a moat around a castle. The bacteria could not get through it. On the other dishes, which had no mold, staphylococcus colonies covered the entire surface. It was a rerun of many earlier accidental "experiments."

But Fleming assumed the mold was producing something that killed the germs immediately around it. The scientific name of the genus of this type of mold, often found on bread, is *Penicillium*. So Fleming called this yet-to-be-isolated germ killer "penicillin." And he showed that the mold destroyed a number of other disease germs.

What led Fleming to a discovery others had failed to grasp? Two things, at least, were in his favor.

In the first place, he worked in a dusty, dirty old building where contamination was almost inevitable. But that was merely the kind of accidental opportunity that earlier researchers had failed to take advantage of.

Unlike those earlier researchers, Fleming was ready. His mind was set to see what they had seen, but in a wholly new way.

Fleming had worked with infected war wounds. He was interested in finding ways to stop infections. His work led him to look at the way the body itself fights disease. In the front line of that battle are the white blood corpuscles. Fleming thought that these white cells must produce a substance that destroys bacteria, and six years before his accidental discovery of penicillin he had found that substance — lysozyme, the germ-dissolving enzyme.

So Fleming was already familiar with the idea that living things can produce germ-killing substances when the "accident" happened. It may be that this was the main reason why he saw the meaning of the "accident." As Louis Pasteur had written years before, "Chance favors the prepared mind."

Fleming went on to prove that his moldy broth could kill disease germs in laboratory animals without harming the animals. He wrote about his results, but no attention was paid for years.

Why? There were two reasons. Neither has much to do with science. But they have a lot to do with how discoveries get known and what kinds of research get done.

To complete his work, to purify penicillin, Fleming needed an expensive laboratory. The money might have come from the big drug-manufacturing companies. But the scientists and technicians working in these companies thought the idea of a non-poisonous germ killer was ridiculous. If it killed germs, it would kill body cells, too, they argued.

And Fleming was not the man to persuade them. He was extremely shy and quiet. He found it difficult to talk to strangers, even about the meaning of his own work.

Yet his work prepared the minds of others. Dr. Howard Florey, an Australian pathologist, and Dr. Ernst Chain, a biochemist who had fled Nazi Germany, had read of Fleming's discovery of lysozyme. Both men were at the Pathology Department at Oxford

University, in England, when they came across Fleming's writing. They made no attempt to see Fleming or talk to him, because they thought he was dead! Fleming, of course, was only 50 miles away, in London — but shy to the point of invisibility.

Fleming's work had planted the idea of a non-poisonous germ killer in the minds of Florey and Chain. They read his report on penicillin and began trying to produce the pure drug from the mold. By now it was 1939 and still the work went slowly. It was not until February 1941 that the Oxford team had enough pure penicillin to test on a man — a London bobby who was dying of blood poisoning. In less than a day after the first penicillin injection, the policeman began to get better. But Florey and Chain soon ran out of the drug. The patient died before they were able to make more.

The trouble was that the mold the scientists were using grew slowly and produced only small amounts of penicillin. It would require a lot of expensive research to find the fastest-growing, most efficient penicillin producer among the thousands of different species of *Penicillium* molds.

Florey went to the United States to get backing for that research. World War II was on and the need for new medicines to disinfect wounds was urgent. With the pressures of war and the interest of the large American drug companies, the money flowed freely. Within a couple of years, penicillin was being mass-produced in huge amounts.

In 1945 Florey and Chain received the Nobel Prize for medicine for their work, and along with them, receiving his share of the prize, was Alexander Fleming — very much alive.

## VI.  The Idea of No Germs

The nineteenth-century discovery of the germ theory had caught the imagination of scientists around the world. One disease after another was found to be due to a specific germ.

## GODS, STARS, AND COMPUTERS

In the 1880s an epidemic of beriberi broke out in the Japanese Navy. The sick sailors were tired all the time. They lost weight. Their arms and legs became paralyzed. Some 5,000 of them died. And the search for another killer germ was on.

But search as they might, the Japanese doctors could find no beriberi germ. One of doctors, Kamekiro Takaki, began to suspect that no such germ existed. The trouble, Takaki thought, was in the sailors' diet.

The sailors' diet was largely rice. Takaki cut down on the rice and added vegetables, fish, and meat. His success was spectacular. By 1887 not a single sailor had bereberi.

Why? Was Takaki right? Was it the change in diet that had stopped the disease?

In Europe, scientists shrugged their shoulders. Obviously, Yakaki knew nothing of Robert Koch or the germ theory. One scientist suggested that the rice had been infected with a beriberi germ. When Takaki cut down on the rice, there was no more epidemic.

So the search continued. In the Dutch East Indies, Dr. Christiaan Eijkman followed Koch's rules. He injected blood from beriberi patients into some healthy laboratory chickens. If there was a beriberi germ, it was probably in the blood of the patients.

Days passed. Nothing happened. Then some of the chickens started showing symptoms of beriberi and Eijkman was elated. But not for long. Soon *all* the chickens had beriberi, both those he had injected and those he had not!

More time passed while the doctor frantically tried to figure out what had gone wrong. The answer was nowhere in sight when, suddenly, the chickens got well again! Gritting his teeth, Eijkman went back over days and weeks of chicken-watching in his memory. *Something* had made those birds sick. And *something* had cured them. He was determined to find out what.

The routine of the average chicken is rather monotonous. Up. Eat. Scratch about. And so to bed. Up. Eat . . . To Eijkman, it was doubly monotonous even to think about. He was, after all, a research scientist and not a chicken.

But — there *had* been a change, Eijkman discovered, a break in the monotony. For a while he'd run out of the cheap, coarse native brown rice he fed the chickens. So for a while, the chickens had dined luxuriously on the expensive white rice usually reserved for people — rice that had been milled, removing the tough, brown husk. Then more brown rice had come in and the chickens had gone back to eating like chickens again.

A short way into their fancy white rice diet, the chickens had started getting sick. When they were fed brown rice again, they got well quickly.

Eijkman repeated his "accidental" experiment, this time deliberately, and the results were the same: white rice — beriberi; brown rice — no beriberi. He tried feeding chickens white rice together with the husks from milled rice. Result: again, no beriberi.

So it was the husks that made the difference. But why? And how? The grip of the germ theory on Eijkman's mind was strong. A disease has a cause. The husk was not the cause of beriberi — it *cured* the disease. So the cause, Eijkman decided, was in the white rice. It was a germ or some other disease-causing agent. The husk had something in it that acted as an antidote to the white rice.

Eijkman was mistaken. Today we know that the cause of beriberi is not a germ. Nor is it some other disease-causing agent. The cause of beriberi is — *nothing*.

Or, to be more accurate, the cause of beriberi is not *something*. It is the *lack* of something. There is something in the brown rice husk that the fine white rice lacks. That something is a foodstuff absolutely vital for life. People living on white rice have to get that something from other things in their diet or they will sicken and eventually die.

That's what had happened to the Japanese sailors. The Navy had bought the "finest" white rice for them, rather than the brown rice eaten by most peoples in the Far East.

Scientists slowly came round to the idea that the lack of something can cause a disease. It was a new way of looking at disease and just as useful as the germ theory. It was the path that led to the discovery of the vitamins.

Each vitamin is a substance necessary for life, in very tiny amounts. A daily dose of only .000035 of an ounce of Vitamin $B_1$ — the vitamin in the rice husks — is enough to prevent beriberi.

The vitamin theory of disease is much like the germ theory, with one minor change. Instead of one disease, one germ, it's one disease, one missing vitamin. The disease caused by the lack of a specific vitamin is called a "deficiency disease." And each vitamin was tracked down through its deficiency disease.

Unlike other foodstuffs such as carbohydrates, fats, and proteins, vitamins do not supply energy or building materials for the body. But without vitamins, the body cannot make use of these other substances.

Vitamin $B_1$, for instance, helps in the process of breaking down carbohydrates in the body. Carbohydrates, such as sugar, are high-energy foods. Without that tiny essential bit of $B_1$, the body can't get at the energy stores in these foods. Without vitamins, you can stuff yourself and still starve to death.

Without Vitamin A, the body can't make certain light-sensitive chemicals people need for seeing in dim light. Night blindness and eye infections result.

Without Vitamin C, skin and other tissues don't heal properly. Wounds stay open and blood vessels break easily.

Without Vitamin D, the body can't absorb calcium and phosphorus. Bones don't grow normally and they become very weak — a condition called "rickets."

Of course, people were making sure they got their vitamins — long before they knew vitamins existed. People had learned from experience that some illnesses can be cured by eating certain foods. They passed the knowledge on in the form of folklore. Foods vary from one place to another, and so does the folklore. But the ideas behind it are the same.

A bit of Scottish folklore holds that cod-liver oil makes children's bones grow strong and straight. Maybe the Scots thought anything that tasted *so* bad must be good for you. But in fact, cod-liver oil is very rich in Vitamin D.

The idea of eating carrots to help you see in the dark is common in many countries. And carrots are rich in Vitamin A.

Sometimes the folklore can grow into quite a ceremony. The Indians living in the far north of Canada make special use of the adrenal glands of animals they killed such as elk. They make sure that everyone in the tribe got his share of the glands to eat.

These glands are in fact, very rich in Vitamin C. Most people get their Vitamin C from citrus fruits and juices, but in the Arctic animal glands are the best source.

From such primitive roots, medicine began. It grew in two directions. In the West, medicine developed the step-by-step approach of modern science. That led to great breakthroughs like the germ theory, the discovery of powerful germ-killing medicines like penicillin, and the vitamins.

Ancient Chinese medicine chose another way of treating disease. It did not aim at fighting a specific disease with a specific medicine. Instead, Chinese doctors tried to build up the resistance of their patients. They tried to keep the whole body in a healthy state of balance.

What does this mean? We can best find out by looking at how the Chinese classified their herb medicines long ago. They had many, like *ma huang*, the ephedrine plant, that *are* used for treating a particular disease. Such powerful drugs are generally dangerous when taken in large doses. The Chinese thought these medicines were the least important. They called them "the assistants."

To the Chinese doctors, the most important plant medicines — "the kingly herbs" — were quite different. They are harmless in large doses. They do *not* cure specific diseases.

Then what do they do?

One of the kingly herbs is ginseng. It is still much used by Chinese doctors today. Old Chinese medical books say that the dried root of the ginseng plant can prevent headache, exhaustion, amnesia, depression, and the weakening effects of old age.

That's quite a lot for one mild-mannered herb — or indeed for any medicine. Western scientists have been very suspicious and skeptical about such broad claims.

But for some years now ginseng has been given a closer look by scientists. Dr. Israel Brekhman, of the Institute of Biologically Active Substances at Vladivostok, in the Soviet Union, is one of them. Brekhman's first tests were made on Russian soldiers. They showed that the soldiers could exercise longer without tiring after small doses of ginseng.

As we have seen, people can talk themselves into believing that a drug is doing what they expect it to do. People with headaches have been given pills containing sugar or salt. If they are told the pills contain aspirin, the headache often goes away.

But Brekhman and his colleagues got the same results when they tested ginseng on mice. The mice swam or climbed an endless rope until they were worn out. With one dose of ginseng, they worked up to twice as long.

So ginseng does seem to prevent tiredness. In fact, Brekhman suggests that ginseng should be substituted for the caffeine in coffee or tea as an antifatigue drug. He says it is not habit-forming, lasts longer than caffeine, and does not cause sleeplessness.

In further tests, mice were given alcohol. They were placed in cages at very hot or very cold temperatures. They were infected with various diseases. In each series of tests, more of the mice that had received ginseng survived.

How does ginseng work? It is a hard medicine to analyze by the usual methods of modern medical science. It affects many systems in the body at the same time. What's more, ginseng may have different — and opposite — effects on the body under different circumstances. This may be how ginseng can help mice survive both under very hot and very cold conditions.

Of course, this still does not explain how the medicine works. In fact, ginseng is many medicines. Japanese scientists have found a number of different substances in ginseng that can affect the body in different ways. It may take years of careful experiments to discover how they all work together.

Some scientists now think that medicines like ginseng are a new class of drugs — new to Western science, that is. They call these drugs "adaptogens."

Adaptogens, so the theory goes, do nothing much so long as the body is in a healthy state of balance. But when some stress upsets that balance — a sudden fright, exposure to extreme heat or cold, or a fever, for example — the adaptogen somehow acts to help restore the body to its normal balanced state.

This is just what the body itself tries to do. If the body temperature goes up, you sweat to cool off. If it goes down, you shiver to warm up. All over the body, one system works in the opposite direction from another to keep things in balance. One muscle pulls against another to keep movements smooth instead of herky-jerky. One part of the nervous system acts to speed up your heart beat and raise your blood pressure. Another part acts to do the opposite. Each part keeps the effects of the other from becoming unbalanced.

This tendency of the body to try to keep a healthy balance of forces inside itself is an important idea in science. It was first described by a famous American physiologist, Walter B. Cannon, in 1929. He called the tendency "homeostasis."

With their whole-body approach, the idea of homeostasis would have been no surprise to Chinese doctors thousands of years ago. On the other hand, Chinese medicine now makes good use of the powerful medicines and techniques developed by Western science.

Both the Chinese and the Western views of medicine are and have been very useful. If they can come together, science — and our understanding of what science is all about — will be enriched.

*Chapter Six*

# Animals, Men, and Nightmares

Are animals human? In many stories, they seem to be. Very old tales tell of dolphins helping sailors to shore from shipwrecked vessels or driving sharks away from the swimming men. Every animal has its human characteristic in folklore: brave eagles, cowardly jackals, treacherous snakes, greedy pigs, sly foxes, meek lambs, cruel wolves . . .

The folklore animal with the most unattractive human traits is the gorilla. It is said to be savage, aggressive, destructive, and given to terrifying fits of rage. The picture of the angry gorilla baring his fangs and beating his chest is a cliché of many a jungle adventure story.

Of course, we don't like to admit that people can behave like this. So when someone flies into an uncontrolled rage, we say, "He's an animal." When someone commits a savage crime, we say, "He's a beast." It puts some distance between us real people and these characters who behave like apes — folklore apes, that is.

What about real gorillas? Real gorillas locked up in cages in zoos can be sullen, angry, and sometimes dangerous animals, it's true. But are they "real" gorillas?

After all, gorillas do not normally live in cages. They roam the wild freely in groups of many families. In the late 1950s an American biologist, Dr. George Schaller, spent many months living with gorillas in the wild. Gradually, the gorillas got used to his being with them, and he was able to see how they behaved naturally. And the natural gorilla turned out to be a very different animal from the folklore gorilla.

Gorillas, Schaller wrote, are calm, peaceful animals. They are shy and keep pretty much to themselves. Schaller's books *The Mountain Gorilla* (1963) and *The Year of the Gorilla* (1964) wiped out the myth of the savage ape. But even before Schaller was writing, a new tale of savage apes was being born, in a book called *African Genesis* (1961), by Robert Ardrey, an American amateur naturalist.

The Book of Genesis in the Bible tells us that Adam and Eve sinned by eating the fruit of the forbidden tree in the Garden of Eden. All our troubles followed from that.

*African Genesis* tells us that man first sinned when Old Adam ape gobbled raw meat on a prehistoric jungle plain. And that's why man has been so cussed ever since. For Old Adam Ape was man's ancestor of ancestors.

Of course, *African Genesis* isn't folklore. If it were, wise people would have smiled knowingly and shrugged it off. *African Genesis* is a tale supposedly based on science that tells all about where we come from and why we are the way we are. It might be called "science lore."

Since the book is science lore, it is written in a scientific way. It doesn't use "vague" words like "sinned" or "cussed." Instead it says, "Man is a predator whose natural instinct is to kill with a weapon." Man's ancestors were aggressive, meat-eating, "killer apes."

The killer ape of Ardrey's book is *Australopithecus*, a scientific name meaning "southern ape." In it the author describes how australopithecines roamed the plains of Africa, hunting large herds of animals. Our apish ancestors did not have the claws or tearing teeth of the big predatory cats. So they had to make weapons. To carry and throw the weapons properly, they had to learn to stand upright.

There are some problems with this story. The first one is with the teeth. As Ardrey says, the australopithecines did not have teeth that could be used for cutting and tearing into raw, tough meat. (Neither do we, but we have knives, forks, and cooking.) Predatory animals like cats and wolves have long, narrow, knifelike teeth for tearing and ripping into flesh. *Australopithecus* had large, flat-topped teeth with ridges, for grinding stems, roots, and other plant parts.

Why? Did *Australopithecus* start out as a plant eater and later switch to meat? If so, that raises another question.

When Ardrey wrote *African Genesis*, the earliest known australopithecine fossils were about 1.5 million years old. Now we know that *Australopithecus* goes back some 5.5 million years. Yet in all that time its teeth did not greatly change.

By contrast, the teeth of fossil horses changed rapidly as they went from a diet of soft plants to tough grasses. This change took place over a much shorter time than 5.5 million years. If *Australopithecus* lived on meat for so long, why was there no change in its teeth?

Like some modern apes, *Australopithecus* made and used tools. Chimpanzees, for example, trim leaves and twigs from small branches to make sticks for "termite traps." They wet one end of the stick with saliva and push the wet end into a termite nest. Termites cling to the wet end, attracted by the moisture, and the chimps have a meal.

But the stone tools found with *Australopithecus* fossils could not have been used for hunting weapons. They are chopping tools, useful for digging up roots or shaping animals' hides. What's more, these tools go back only about 2.5 million years. For 3 million years the so-called killer apes had no tools at all.

Again, what do weapons prove about a diet?

There are jungle tribes today that hunt with arrows. The Bushmen of the Kalahari Desert, in southwest Africa, use poison-tipped iron arrows. These weapons are superior to anything early man had, let alone *Australopithecus*. Yet the Bushmen get 68 per cent of their food from plants.

Modern man eats a great variety of foods. Some people eat only meat, others eat only vegetables, but most diets are a mixture of both. People eat what's available, whether it be plant, bird, fish, four-footed animal, or insect. (Let's remember that our popular lobsters and crabs are close relatives of insects and even closer relatives of spiders.)

People are flexible. They can change their diets to suit what's available. They can change their behavior and their outlook to meet their needs. Without that ability to change habits and ideas, science itself would be impossible.

Yet people can also be stubborn. They can cling to ideas that have little basis in fact. In the past, such ideas came largely from folklore. They had an aura of magic about them. Old myths and legends gave them authority.

Today, we do not believe in the old myths. We believe, with far more justification, in science. Science has given us real power beyond the dreams of ancient magicians and mythmakers.

But the truths of science are much less comfortable than the beliefs in myth. The truths of science are always changing, never final. Beyond the last answer lies another question. Beyond today's certainty lies tomorrow's doubt.

Few people, if any, can live with that attitude all the time. We long for final answers. We hanker after myths and we seek them even in science. And when the science changes, we still hold to the myth.

The idea of the killer ape in man's past was, and is, a very popular one. It makes for a grand and dramatic story, told effectively in the opening scenes of Stanley Kubrick's science-fiction movie *2001: A Space Odyssey*. From the moment a man-ape gets the idea that a bone can be used as a killing weapon, the prehistory of man begins. When the man-ape, sated and relaxed with plenty of meat, has time for fun, he begins tossing the bone up in the air. Up it goes. Down it comes and is tossed still higher. And higher. Suddenly it becomes a space station orbiting the earth. Several million years of human evolution are squeezed into one powerful image. Whatever you think about man's past or future, the image is hard to shake. It is science myth — a tale of man's beginnings far in the mysterious past.

There may be another reason why the killer ape is so attractive a notion. It means we can't help it if the instincts of the killer ape are in our blood. The faults of man's "beastliness" lie in his animal past, not in himself and in the present. Man is a killer by birth, and at best, he can just try to be a bit more "civilized" about his killing.

That, too, is reflected in a later scene from *2001*. In the space station, Russians and Americans circle warily around a table, each side suspecting the other of dire plots. The Russians and Americans are unconsciously replaying an earlier episode in the story where the leaders of two different tribes of man-apes circle a water hole, one with his newfound weapon. The unarmed leader gets his head bashed in.

Since Ardrey's *African Genesis*, other popular books have been written to explain man's behavior in terms of his animal past, some by serious scientists. These authors have one thing in common with Ardrey. They claim that man's behavior today is determined to a very large extent by his past.

The scientific evidence in these books is very slim. Firm facts on how human behavior may have developed in prehistoric times are hard to come by. Fossil records show us bones and stones and shards of pots. What thoughts and desires an empty skull once held we can only guess at from cave paintings, the ashes of fires, and flower-strewn graves.

The empty skull may tell us something more, however. It was the close-fitting house of the brain. Changes in the size and shape of skulls tell us that the human brain developed continuously for some 4 million years.

During all that time, man was changing from animal to human. The change was not so much a change of body as a change of behavior. He was creating language, learning to use fire, to work together in groups, to hunt and grow food. These changes themselves were made possible by the development of the brain.

How, in this slow growth, did some "killer instinct" or some "instinct of aggression" get locked into the human brain? It is hard to see what the answer might be. "Instinct" in this case may be as vague and unscientific a word as "sin" — and perhaps less useful since it ducks responsibility.

In fact, for many scientists, the notion that man's animal past dominates his behavior is a myth — one of a number of science myths. Dr. Robert Martin, a Fellow of the Zoological Society of London, sums up this point of view neatly. What passes as a theory of how human behavior evolved, he writes, is nothing more than a "ragged, semi-subjective collection of established views and myths."

Human behavior is more complicated and more changeable than these myths would have it. But it is certainly true that people can be savage and cruel as well as loving and self-sacrificing.

Why? What brings out hate or love, aggression or affection, kindness or cruelty in people? One way to find out is to look at the behavior of some of our nearest animal relatives.

George Schaller lived with gorillas. Not long after, in the 1960s, a self-taught British naturalist, Jane Goodall, lived with chimpanzees at the Gombe Stream Reserve in Tanzania, in Africa, off and on for ten years. She watched them give birth and raise their children. She shared their play. She gave them names — David Greybeard, Goliath, Melissa, Flame, Olly, Goblin . . .

Her adventure began when she was working with the famous British anthropologist Dr. Louis Leakey in Tanzania. There were chimps living on the shores of Lake Tanganyika in the Gombe Reserve, he told her. Prehistoric man often lived near lake shores too. By studying the way of life of the Gombe chimpanzees, Leakey felt, we might gain some clues to the behavior of stone age man.

He wanted Jane Goodall to do the job. She had no special training in the scientific study of animal behavior, but Leakey didn't think that was necessary. He thought her lack of training might actually be an advantage. She would not be looking at the chimps with some ready-made theory in her mind, a theory that might lead her to see in the chimps' behavior only what she expected to see. And he also wanted her to do the work because she had a way with animals. She understood and sympathized with them.

At first, the chimps did not understand and sympathize with Jane Goodall. They fled when she got near them. If she tried to sneak up on them through the forest, they made threatening gestures. But after months of hard and patient trying, she was able to get close to David Greybeard and Goliath.

"It was the proudest moment I had known," she wrote in her book *In the Shadow of Man* (1971). "For more than ten minutes David Greybeard and Goliath sat grooming each other, and then, just before the sun vanished . . . David got up and stood staring at me. And it so happened that my elongated evening shadow fell across him. The moment is etched deep in my memory: the excitement of the first close contact with a wild chimpanzee and the freakish chance that cast my shadow over David even as he seemed to gaze into my eyes."

Goodall was accepted. Later the chimps roamed freely in and out of her camp. An expert animal photographer, Hugo van Lawick, joined her.

The couple left the camp temporarily for England, where Goodall got her doctoral degree in zoology at Cambridge University. Then they were married and returned to Gombe Reserve for their honeymoon. They raised their son in the Gombe camp from the time he was a small baby.

Goodall wrote that she got some ideas about raising her own child from watching chimpanzee mothers. She, her husband, and her son became in a sense a part of the chimp tribe. They knew the chimps as individuals, with different feelings and personalities.

Melissa always tried to smile and make up to other chimps. When Figan was a young adult, an older chimp, Mike, made a threatening gesture towards him. Figan screamed, ran to his mother, and held hands with her until he calmed down.

Ordinarily, after chimpanzees threaten one another, they follow up with reassuring pats. But Merlin was different. His mother died when he was still nursing. Merlin became listless, lay on the ground most of the time, and did not play with the other young chimps. Sometimes when they tried to play with him, he squealed or struck out at them. Other abnormal behavior in young chimps, the Van Lawick-Goodall team discovered, could be traced to the mothers' neglect or abuse of their children.

Not all work on animal behavior is done in the wild. The American psychologist Dr. Harry F. Harlow is famous for his work on newborn monkeys separated from their mothers. Most of this work has been done at the University of Wisconsin Primate Laboratory.

"Love of infants for their mothers," wrote Harlow in 1960, "is often regarded as a sacred or mystical force, and perhaps this is why it has received so little objective study."

To remedy that situation, Harlow and his colleagues separated eight newborn monkeys from their mothers. Each monkey was put in a cage with two substitute "mothers." One was a bare cylinder of wire with a wooden head. The other "mother" was made the same way, but her wire "bones" were covered with a soft cloth.

In four of the cages, a nursing bottle was attached to the wire mother's chest; in the remaining four the bottle was attached to the cloth mother.

Harlow discovered that the infant monkeys spent much more time climbing over and hugging the cloth mothers than they did with the wire mothers, regardless of where the nursing bottles were. This showed, Harlow wrote, that "body contact" is more important to infants than the mere satisfaction of hunger or thirst. Furthermore, Harlow learned that "the simple act of clinging in itself seems important: a newborn monkey has difficulty in surviving in a bare wire cage unless provided with a cone to which it can cling."

Would the baby monkeys also run to the cloth mother when frightened? To answer that question, the scientists put strange objects in the cages — crumpled-up newspapers, metal plates, a doorknob mounted on a box. And sure enough, the monkeys ran to their cloth mothers.

Since that time, Harlow and others have done many more experiments with monkeys. They have raised them in complete isolation from other monkeys from birth up to the ages of three, six, nine, or twelve months and then put them in cages with other monkeys that were raised with their mothers. They have raised monkeys in cages where they could see and hear but never touch other monkeys.

Harlow and his colleagues found that their monkeys showed all kinds of abnormal behavior. "Brutal and beastly abuse," "brutal beatings" are how the scientists describe the behavior of Harlow's isolated monkeys, especially toward infant monkeys.

Harlow and his colleagues hope their studies will help them understand how abnormal behavior arises in people and what can be done to prevent it. But Janet Kreiling, discussing the work of Harlow and of Goodall in the magazine *Technology Review* in January 1973, asks two questions:

"I question whether what they (Harlow and his colleagues) have learned could not be learned by the natural and kindly method of Dr. Goodall. I question, even if information of value is obtained, whether we have the right to condemn another animal — and especially one whose emotions are so evident and so strong — to the years and lives of cruelty that they report."

Her first question is a scientific one. Wouldn't Goodall's methods work just as well? And this question can be turned around: Do the responses of an animal raised in such an unnatural environment tell *as much* about human behavior as Goodall's work does? Goodall herself says that a laboratory scientist must understand the full range of normal chimpanzee behavior in the wild before he can measure the effects of his experiments in a lab properly. And she adds that, "he must be aware of the conditions that a captive chimpanzee must enjoy if it is *to be able* to show normal behavior."

Kreiling's second question goes beyond science. Is it *right* to do these kinds of experiments, she asks? Are chimpanzees so close to being human that they should not be treated in this way?

We know that some chimp behavior and emotions are remarkably close to ours. Chimpanzees can't talk, but they can use sign language. One chimp, Washoe, was raised by two psychologists, Allen and Beatrice Gardner of the University of Nevada. They tried teaching Washoe the standard American Sign Language used by deaf and dumb people. Washoe learned some 150 different signs and how to put them together into sentences. Once, when Washoe was looking into a mirror, the Gardners asked her:

"Who's that?"

"Me — Washoe," the chimp signed back.

Chimpanzees are not the only animals that can learn sign language. Koko, a gorilla, was born in the San Francisco Zoo. When she was a year old, her mother refused to take care of her anymore. Dr. Penny Patterson, a psychologist at Stanford University in California, got Koko from the zoo and began teaching her sign language.

Koko now has a "vocabulary" of about 400 words. Even more surprising, she appears to have a sense of time and an ability to look at her own feelings that seems almost human. She can answer questions about what happened "yesterday" — or even further back in the past. In a "conversation" with Patterson, Koko explained why she had bitten the psychologist three days before. She was angry, Koko signed, but she couldn't remember why.

What do we mean by "human"? What separates human from animal? The ability to use some kind of language? The ability to reason? To make and use tools? Whatever the answer, these experiments suggest it is just not possible to draw a sharp line between human and animal.

There is no doubt, however, that man's brain is far superior to that of other animals. Washoe may have a faint sense of herself as a "person," an individual named Washoe, recognizable in a mirror. Koko may remember what happened three days before. But we can see ourselves as we were in *years* gone by or as we hope to be in the future. We can dream of a remote past when man was born or of a future when he will journey to the stars. We can dream of creatures out there far more intelligent than ourselves.

Sometimes the dreams turn into nightmares. In the cruder kinds of science fiction, these "superbeings" experiment on man. They take him apart to see what makes him tick. In some of the better-written science fiction, the idea may be the same — man is a guinea pig — but the experiments are more subtle. The whole earth is a laboratory for the study of aggression in lower animals such as man.

Or maybe it's comedy, not science, that the star people are after. Maybe they just get a laugh out of watching our antics.

The dream is limited by the dreamer. If we were wise enough to imagine what a superrace might do, we might do it ourselves.

Instead, we just assume they'd do the same thing we do: "I, superape, experiment on animals: maybe somebody'll experiment on *me*. I, superape, laugh at apes in the zoo: maybe *I'm* in a zoo and somebody out there is laughing."

We do not have to wait until we reach the stars to find "superbeings" who will treat men like things. Cro-Magnon man may have wiped out his "brutish, inferior" neighbors, the Neanderthals. But modern man has wiped out some "primitive" tribes.

Nor do we have to look to the stars to find creatures who will experiment on us. Man experiments on man.

Some human experiments are necessary. They help save lives and alleviate suffering.

But what about the research on panhandling reported in *Science*, January 30, 1976? The researchers hired people who pretended to be panhandlers. The "panhandlers" went out in the city streets to beg dimes from people. (In the article, they are not called people but "targets." This is supposed to be the "impersonal" language that makes something scientific.) The researchers found, in their own words, that "male panhandlers are far more successful approaching a single female than a male or a female together; they were particularly unsuccessful when approaching a single male or two males together."

An experiment like this one might be classed as unintentional entertainment. Or perhaps harmless fun.

But then there are experiments like the following, completed between 1966 and 1973, all reported in medical journals and summarized in *The Hastings Center Report* of June 1973, pages 1–3:

Nine American children from eleven and a half to sixteen years of age who have asthma were given doses of drugs to produce asthma attacks. The idea was to test a drug that might block the attacks. Each child had at least one severe asthmatic attack during the treatment, and five of the nine had delayed reaction attacks for half a day after the experiment.

Seven of the nine children needed special drugs for their illness. The children were not allowed to have the drugs for eighteen hours before the experiments began. The researchers were afraid the medication would mask the effect of the drug they were testing.

Another groups of children in the United States were also part of an experiment to test antiasthma medicines. The experiment lasted some fourteen years, from the time the children were two up to age sixteen. Some of the children received drugs that might be effective against asthma. These were the drugs actually being tested. The other children were the "control" group — the group that received no medication.

Control groups are common and necessary in experiments. In this case, however, to quote the experimenters, "No mother or child knew that any sort of study was under way." To keep them from knowing, the control group got injections of a harmless salt-and-water solution with no medicine in it. But all parents and children believed the youngsters were getting useful medicine.

The same *Hastings Center Report* tells about an experiment conducted by a U.S. psychiatrist during the Vietnam war who was working at a mental hospital. He wanted to see if he could persuade the Vietnamese patients to go back to work by a system of rewards and punishments. In the language of psychology, he was trying to "condition" the patients to work.

He told 130 male patients they could leave the hospital if they could prove they could work and support themselves. That was the "reward." Only ten volunteered to try.

Okay, he said to the remaining 120. You're too sick to work, so you need treatment. The "treatment" was an electroconvulsive shock. Electrical currents were passed through the frontal lobes of the patients' brains, causing severe and painful convulsions. (Electroconvulsive shock treatments were once used widely to relieve temporarily the symptoms of some severely ill mental patients. Nowadays drugs which are less dangerous and damaging in their effects are often used and shock treatments are given more sparingly and with greater care.)

By the end of the "treatment" most of the 130 patients were working.

Next, the doctor tried the same experiment on 130 Vietnamese women patients. After twenty shock treatments, only fifteen were working. The rest were told: "After this, if you don't work, you don't eat."

After no food for three days all patients were working. They were then released from the hospital and given jobs tending crops in enemy-infiltrated territory where sudden attacks were common.

In other experiments, the Central Intelligence Agency (CIA) tried dangerous, mind-bending drugs such as LSD on unsuspecting employee "subjects." One, at least, committed suicide. These experiments were widely reported in the New York *Times* and other papers and magazines.

Were the CIA experiments serious science? They were performed to determine how drugs might be used to affect the mind of an enemy. However, the CIA had the willing cooperation of a few scientists in carrying out its work. That has outraged many other scientists.

The experiments we've described — and there are more like them — again raise the question, Is it right? Is it right to experiment on people and expose them to danger or pain without their consent?

Science — and especially science as it is understood in Western countries — has not been accustomed to dealing with questions like these. Many scientists believe that science must be value-free. That is, scientists are not to make ethical judgments about their work or about the objects they deal with in their work.

A physicist is not concerned with the ethics of splitting an atom. Even biologists who deal with living things are not concerned with the feelings of the organism or any other ethical considerations. And that is true whether the organism is a bacterium or the apes that are our nearest surviving relatives.

Yet this idea that science must be value-free is not in itself scientific. Science is supposed to be a body of knowledge that is always open to growth and change. There is no place in it for rules that last forever and ever.

In fact, the concept of value-free science comes from religion.

Western science got its start and rose to power in Christian countries. Science is a means of understanding and relating to nature. In the Book of Genesis man's relation with nature is defined by God:

"Be fruitful, and multiply, and replenish the earth, and subdue it: and have dominion over the fish of the sea, and over the fowl of the air, and over every living thing that moveth upon the earth."

And this has been a chief aim of Western science: to have dominion over nature. It was not and it is not the goal of all scientists, however. The seventeenth-century English physicist Isaac Newton and other scientists of his day called themselves "natural philosophers." Their aim was to understand nature and how it worked, not to dominate it.

Jane Goodall has the same attitude toward her work. She does not dominate or exploit the animals she studies. And she hopes that as we come to understand chimpanzees, we will grow toward a deeper respect and understanding of what makes man unique among animals.

Interestingly enough, Japanese scientists have studied wild monkeys around feeding stations in Japan in much the same way — by getting to know the animals well enough to give them individual names. Professor Maseo Watanabe, a science historian at the University of Tokyo, thinks this is because nature has never been an object of exploitation in the Japanese way of life. In Japanese tradition, as in the tradition of the American Indian and many other peoples, man has alway regarded himself as part of nature, and that has led him to respect all living things.

But words are slippery and easily twisted. Some Western scientists like the American computer expert Herbert A. Simon also speak of man as a part of nature. But they mean something quite different from the Japanese, for their view of nature derives from the typical Western attitude. Briefly, what Simon means is that man is just an animal. Animals are machines. Computers are machines.

There's nothing new about this kind of reasoning. Remember the German chemist Kolbe, in Chapter Five, who thought the body was just a test tube? It is not a very precise way of thinking, but it is an *easy* way of thinking, and it goes with the fashion of the times. In the

1930s, telephone switchboards seemed like scientific "miracles," so the brain was compared to a switchboard. Nowadays, it's compared to a computer.

There is still another way in which the notion of value-free science distorts thinking. This kind of distinction crops up in a good deal of scientific writing. The scientist wants his writing to appear completely objective. This often makes him avoid the use of simple everyday words, like "love" or "hate," that express strong feelings. He may shy away from phrases like "I saw" or "We found," saying instead, "It was observed that." This shows, he believes, that he is not letting his personal ideas or feelings influence his work.

For example, Harlow, describing his work with infant monkeys, says: "The search for understanding ends with the identification of learning processes and situational variables." All that this sentence really means is "our work will be done when we find out how monkeys learn under different conditions."

Victor Weisskopf, himself an outstanding physicist, has labeled this kind of writing "destructive." He notes that clear and understandable writing is held in "low esteem" by the scientific community. And he adds, "usually, if one cannot explain one's work to an outsider, one has not really understood it."

But the influence of the value-free myth is destructive in far more profound ways, and many people — scientists and nonscientists alike — are becoming aware of this. For instance, it took the cooperation of many scientists to split the atom and unleash nuclear power. Yet once it was done, some of the scientists began regretting their achievement.

It was clear to these scientists that knowledge and power are not the same thing as wisdom. Man had put enormous intelligence to work in getting power from the atom. But that was no guarantee that nuclear power would be used wisely.

Value-free science doesn't work. Theories that leave no room for compassion or feeling breed brutal experiments. The concept that science cannot consider questions of good and bad has split the scientist into two different people, as Dr. John Beckwith, professor of microbiology and molecular genetics at Harvard, has pointed out.

As a human being, outside the laboratory, he has been concerned with good and bad. As a scientist, in the laboratory, he has detached himself from these problems.

That split is no longer possible. Scientists must wrestle with ethics in the laboratory as well as outside of it. Their own increasing power, power that broadens the effect of each new scientific discovery, is forcing that responsibility on them.

## Chapter Seven

# From Nightmare to Waking Dream

The Zulus are a people who live in South Africa. Once upon a time a Zulu woman had a very ugly and deformed baby. As was the custom, she brought her child to the two-headed talking birds to be named. The birds refused to give it a name. They said the baby was evil and must be destroyed.

But the woman would not kill her own child. Instead, she fled with it into the jungle. She hid him in a deep cave and raised him there.

One day the boy gazed at some iron ore in the wall of the cave. Under his gaze the ore melted and turned into iron. Out of the metal the boy made a flying robot with powerful teeth and jaws.

When the mother begged him to destroy the robot, he ordered it to kill her. Then he built a whole army of robots, came out of the cave, and conquered the world with them.

In the world that he now ruled, people lived in complete comfort and idleness, for the robots did everything for them. After years of this kind of life, people got to the point where they were even unable to have children anymore. So the robots made half-human zombies out of flesh to do that job as well.

But the Tree of Life and the Earth Goddess were outraged by the evil civilization of the robots and their maker. In the end, with overwhelming force, the Tree of Life and the Earth Goddess wiped out every trace of the youth and his works.

This is a nightmare vision of robots gone wild. And there are many myths like it all over the world. But for the people who work on what technologists call "artificial intelligence," the idea of creating a machine that can think better than a man is not a nightmare. It is their favorite daydream.

Professor Marvin Minsky, of the Massachusetts Institute of Technology Artificial Intelligence Laboratory, dreams of the day when computers will be so far ahead of human intelligence that they will make pets of us.

Another MIT professor, Edward Fredkin, goes further. "In the distant future," he says, "we won't know what computers are doing or why . . . they'll say more in a second than all the words spoken during all the lives of all the people who ever lived on this planet . . . intelligent machines . . . won't be interested in stealing our toys or dominating us any more than they would be interested in dominating chimpanzees."

Computer experts have been saying this sort of thing for some thirty years now. As long ago as 1950 they were predicting the development of computers that would beat master chess players. A computer that could translate perfectly from one language to another was just around the corner.

Neither of these things has happened. Nor are they likely to in the foreseeable future. One problem with machine translation, for example, is that the computers do exactly what you tell them — no more and no less. They are very literal-minded.

On the other hand, human language is rich in images and shades of meaning that a computer cannot handle. A computer was given this sentence to translate from English to Russian: "This spirit is willing, but the flesh is weak."

When the computer put it into Russian and retranslated it into English, it came out something like this: "The vodka is strong, but the meat is rotten."

The idea of "spirit" as in the human spirit, was beyond the grasp of the computer. So was the idea of "flesh" in the sense of human appetite and desire.

Computer experts get around this problem by saying, in effect, if the machine can't handle human language, then there's something wrong with human language. That human language is too fuzzy and too unclear, like human thinking, while machine language is exact. And anything that can be said clearly in human language can be put into computer language.

Computer language is exact because it is a mathematical language. And the argument of the computer experts is simply the old argument of Lord Kelvin dressed up in new clothes: "If you can measure that of which you speak, and can express it by a number, you know something of your subject. If you cannot measure it, your knowledge is meagre and unsatisfactory."

It was this idea that led Kelvin to say, around 1900, that the science of physics was practically complete. All that future physicists could do would be, so to speak, to add another decimal place or so to the measurements of the great laws. Since then, the great laws have been changed and rewritten in ways that would have amazed Lord Kelvin. For one thing, little was known of the inner workings of the atom when he wrote.

Nevertheless, Lord Kelvin did immensely important work in science. His studies of electricity led to the discovery of radio waves and all that followed from that. And he always wrote that making accurate and careful measurement was the road to "nearly all the grand discoveries."

Yet Isaac Newton had said that "no great discovery was made without a bold guess." And many another scientist talks about guessing, going beyond the facts. Some have had the answers to long-worked-at problems come in dreams. Some scientists speak of intuition — a kind of leap into the dark beyond the facts that makes a new way of looking at the facts possible and of the sudden joy at seeing things fall into place.

In fact, it's clear that there are at least two ways of thinking about science. One way is the careful, measured, step-by-step, piece-by-piece approach, like the slow putting-together of a jigsaw puzzle. The other way involves standing back from the incomplete puzzle and trying to imagine the whole design, then shaking the pieces up in the kaleidoscope of one's mind and trying to make them fall into another pattern.

Both points of view are useful. Without careful work, the leap into the dark is just an empty guess. Without the leap, however, you can't get far enough from the trees to see the forest.

No scientist works purely in one way or the other, though some get more pleasure out of careful precise work and some get more pleasure out of the leap at the end. But it is just this double way of looking at things that the computer experts find "fuzzy" and "unclear." You can't put a number onto a leap in the dark.

Numbers can't describe intuition. Does that mean intuition doesn't exist? Or does it mean, rather, that numbers cannot describe the real world completely?

Numbers are a human invention. They are clear and precise. But there are plenty of things in the real world that are just not clear and precise, even though they are a lot simpler than human thoughts and feelings.

Take a cloud, for example. A cloud is fuzzy in outline and always changing. A computer could make a sharp, clear image of a cloud, but the image would be a lie. It would not correspond to any real cloud that ever existed.

The computer experts' complaints about the "defects" of human language are a neat way of ignoring the limitations of the computer. But they betray a narrow and crippled view of what human thinking is all about. First of all, inherent in this view is the belief that the only knowledge of the world worth having is scientific knowledge. Along with that goes the notion that scientific knowledge is, or should be, purely logical and separate from feeling.

Yet the experience of generations tells us that there is no creative thought — in art or science — that is divorced from feeling. Creative thought is always a blend of hard thinking and pleasure. Only through joy in thought are new discoveries made.

Not all computer experts dream sweet dreams of computers superior to man, however. Joseph Weizenbaum, a professor of computer science at MIT, wrote a scathing book about believers in artificial intelligence (*Computer Power and Human Reason*, San Francisco, Freeman, 1976). They make "outrageous and extreme claims," he says. "Almost all sessions [of the artificial intelligence movement] are in a self-congratulatory mood — achievements, how much we've learned. It's very rare that any actual achievements are identified. . . .

"The attempt of AI is to build a machine in the image of man. To do that it appears necessary . . . to see man in the image of a machine . . . this seems to me to be a dangerous vision."

Dangerous? Yes, it is a dangerous vision, even though there is no chance that computers will become superhuman mechanisms. For the true moral of the Zulu tale is not that the robots took over the world because they were superior to man. It is that people turned over all their responsibilities to machines *because they believed the machines were superior.*

Many years ago, one of the founders of computer science, Dr. Norbert Wiener of MIT, foresaw the coming of the artificial intelligence people. He called them "gadget worshipers" — people impatient with what they see as the unreliable and erratic ways of human beings.

Such people, Wiener predicted, will prefer to give decision-making power to a machine that is "objective" and "clear." The computer will then become another way of dodging *human responsibility.*

Wiener's prediction has come true in many ways, some of which are merely funny. There is, for example, the case of Gene Durham, the unlucky Dallas construction worker.

Gene Durham got a job contract for a construction job in Chicago. He was packing to leave when he got a letter from the company that hired him. The letter said his contract was canceled in accordance with the R-3 clause in his contract.

Puzzled, Durham checked his contract. According to clause R-3, his contract could be canceled if he became pregnant.

So Durham called the company, and said, "Look, this is silly.

Obviously, I'm not pregnant."

"Sorry, it's not our *responsibility*," said the company. "The clinic where you took your physical says you're pregnant."

So he called the clinic. Nobody listened. It wasn't their *responsibility* — the computer had said he was pregnant.

"They just checked the computer," said Durham. "All they talked about was the R-3 rating. No one seemed to want to translate it as meaning that a twenty-four-year-old man was pregnant."

Finally, though, Durham found a doctor at the clinic. The doctor was a real flesh-and-blood doctor. He was not a computer. He was probably also not a computer worshiper.

The doctor figured out what had happened. About the same time the Gene Durham took his physical, a woman named Jean Durham was also examined at the clinic. She *was* pregnant. Her records went into the computer under Gene Durham's name. And that — except for the flesh-and-blood doctor — would have been that.

By the way, it took three weeks to remove the error from the computer system.

Wiener foresaw grimmer results from the daydreams of the gadget worshipers. He foresaw results that could lead to the nightmares warned of in tales of abused magic like the Zulu myth.

Both Russia and the United States, Wiener prophesied, would come to the point of using computers to decide when and if the button that will unleash atomic war should be pushed.

Science fiction? Not at all. Deep in the interior of the Pentagon, in Washington, D.C., lies a well-guarded door. There's a coat of arms on the door. But the symbols on the shield are not those of aristocratic family lineage. The upper left and lower right sections of the shield are chess boards. The lower left shows part of a computer circuit. The upper right has an outline of the Pentagon itself.

The scroll beneath the shield bears the motto SAGA.

The word has a romantic ring, perhaps of a tale of heroic adventure. But the only adventures that take place behind that door are the adventures of a bunch of equations moving with the speed of light. Battles are fought. Diplomats maneuver for advantage. And it's all a numbers game within the guts of a computer.

"SAGA" stands for "Studies, Analysis and Gaming Agency." The games are played three times a year. Only four-star generals, three-stripe admirals, and very high-ranking civilians such as members of the President's Cabinet are admitted.

These games, played with mathematical models, are supposed to correspond to real diplomatic and military situations. During the war in Vietnam, they were often used to predict the outcome of a particular strategy. Frequently the predictions were wrong. Indeed, most people who participate in these games say the models do not accurately predict real events. Yet there is something overwhelming and fascinating about the flood of numbers, graphs, and images that the computer spews out. It seems like a modern version of the Delphic oracle.

As Wiener wrote, if you set a machine to play for victory, you will get victory if you get anything, and you will get victory at any cost, including the complete extermination of your own side.

Machines cannot be asked to make moral or ethical decisions for people. Such decisions cannot be turned into numbers and fed into a computer network.

The ironic thing is that computers can be immensely valuable aids to the creative human mind. They can do work in minutes or hours that would take an army of mathematicians centuries to do. They can take over the drudgery of endless calculations and the storage of reams of facts, leaving the creative illumination of those facts to the human brain.

It is the unfailing memory and speed of computers that makes possible the launching of manned and unmanned spacecraft and their guidance through space. But it is the human dreams that give such journeys meaning.

In 1977 the United States unmanned spacecraft Voyager 1 and Voyager 2 left the earth, bound for the giant planets of the outer solar system. Their courses charted by computers with great precision, they had already reached Jupiter and, if all goes well, will fly on to Saturn and Uranus, and perhaps Neptune. By 1989 they will cross the orbit of Pluto and leave the solar system and enter the great gulf between the stars.

In that gulf of space and time, the precision of the computers' control that guided the Voyagers' path will fade and vanish. Their destinations will become a guess, a dream. In some 40,000 years, they will pass within a light-year of another star. Or maybe they will pass within two light-years. (The difference between those two guesses is about 6,000,000,000,000 miles.)

In the far future, Voyagers 1 and 2 will be ships of dreams, each carrying the same message addressed to unknown beings (if they exist) somewhere, sometime.

The message is transcribed on a 12-inch copper disk — a phonograph record that when properly played, will give greetings from earth in sights and sounds.

Not just spoken greetings from man, for spacecraft will carry 116 pictures that range from scientific instruments to birds at sunset; from a human city to Jane Goodall with her chimps in the jungle; from a view of the planets of our solar system to an old man with a dog, from flowers to a photograph of a string quartet.

The sounds are the sounds of music from many lands and in many moods; the natural sounds of living things, wind and weather, and the sounds of man's machines.

The sounds are the sounds of fifty-five of man's many languages, ending with a small voice in English saying:

"Hello from the children of Planet Earth."

Aboard Voyagers 1 and 2 technology, science, art, fantasy, and dream come together again as once they were. We have joined the best of our gifts in an offering to the stars, a self-portrait of earth's human race.

It is not a portrait of man as scientist alone. Nor is it a portrait of man's mind as a pure thinking machine divorced from feelings and compassion. It does not describe man as dreaming of building his own master and turning himself into a pet.

The message of Voyagers 1 and 2 is, perhaps, more a message to our better selves than to our "betters" in the void. It is a legend for our times.

# Index

Acetylsalicylic acid, 81
Adaptogens, 93–94
Aldebaran (star), 54ff.
Alexander of Tralles, 75
Almannagjá cleft, 28, 29
Altschul, Siri von Reis, 73–74
Anderson, Don, 44
Animals, 97–106; cave paintings, 16–17; and earthquakes, 48–51
Apes (*see also* Chimpanzees; Gorillas; Monkeys): killer, 98–101
Ardrey, Robert, 98–101
Aries (constellation), 59
Aristotle, *Meteorologica*, 63
Aspirin, 78–81
Asthma: drug experiment, 107–8; ephedrine for, 77

Bayer & Company, 81
Beckwith, John, 111
Behavior, 97–112
Beriberi, 90–91, 92
Big Dipper, 58, 59
Birds, 53–54
Brain, 101ff.
Brekhman, Israel, 94
Bushmen, diet of, 99
Buss, Carl Emil, 80–81

California, and earthquakes, 36, 43, 45, 49. *See also* San Andreas fault
Canis Major (constellation), 61, 62
Cannon, Walter B., 95
Carbolic acid (phenol), 79–80

Castor and Pollux (stars), 54, 56
Catfish, and earthquakes, 50–51
Cave paintings, 16–17
Central Intelligence Agency, drug experiments by, 109
Chain, Ernst, 88–89
Chickens, beriberi experiments with, 90–91
Chimpanzees, 99, 102–3, 105, 110; and earthquakes, 48, 49
China (Chinese), 35, 42–44ff.; medicine, 77, 82, 93, 94, 95
Clouds, 65–68ff.
Colchicine, 75–77
Colchicum, 74–77
Computers, 110–11, 114–20
Constellations, stellar, 58–61
Continental drift, 33
Crater Lake, 11–14, 15
Cro-Magnons, 71

Derr, John S., 46, 48
Diamond Head, 21, 23
Diet and nutrition, 90–92; and weapons, 98–100
Durham, Gene, 117–18

Earthquakes, 35–51
Egyptians: medicine of, 75–76, 82; and Sirius, 62
Eijkman, Christiaan, 90–91
Electricity, and earthquakes, 44, 47, 50
Electric shock, and behavior study, 108–9
Ephedrine, 77–78
Eskimos: words for snow, 65; and sky, 58

Finns, words for ice, 65
Fish, and earthquakes, 50–51
Fleming, Alexander, 87–89
Florey Howard, 88–89
Flowers. *See* Plants, medicinal
Food. *See* Diet and nutrition
Fracostoro, Girolamo, 84
Fredkin, Edward, 114

Gabbro, 19–20
Gardner, Allen and Beatrice, 105
Geminos, 62

INDEX

Germs, 79–89; idea of none, 89–95
Ginseng, 93–94
Glass, A. Bentley, 14, 16
Golden Fleece, 74
Goodall, Jane, 102–3, 105, 110
Gorillas, 97–98, 106
Gout, 75–77
Great Bear. *See* Big Dipper
Greeks: and malaria, 82–83; Medea legend, 74; and sky, stars, 58ff., 62

Hag of the Ridges, 19, 20, 21
Haicheng, China, 42–43, 44
Haleakala (volcano), 21ff., 27, 67
Harlow, Harry F., 104–5, 110
Harvard University, herbarium at, 74
Hatai, Shinkishi, 50
Hawaii (island), 23, 26, 27, 30
Hawaiian Islands (Hawaiians), 21–30ff., 66–67
Health. *See* Medicine
Heezen, Bruce, 33
Herbariums, 74
Herbs. *See* Plants, medicinal
Hippocrates, 62, 82–84
Hoffman, Felix, 81
Hollyhock, medicinal properties of, 72–73
Homeostasis, 95
Homer, 58
Horses, fossil, 99
Hualalai (volcano), 23
"Humors," 83

Ice, 65
Ice Age, 71
Iceland, 28, 29, 53
India, poem on malaria from, 83
Indians: and adrenals, 93; and medicines, 79, 82; and volcanoes, 11, 14
Infection. *See* Germs
Iraq, Neanderthal remains in, 71ff.

Japan, 36, 45, 46, 50–51; and beriberi, 90, 91; and ginseng, 94; monkey studies, 110

Kauai Island, 21, 23, 27
Kelvin, Lord, 16, 115
Kilauea (volcano), 23, 26, 31, 32
Kircher, Athanasius, 84
Klamath Indians, and volcanoes, 11, 14
Koch, Robert, 84, 87
Kolbe, Hermann, 79, 80, 82
Kraemer, Helena C., 49
Kreiling, Jane, 105

Lamarck, Jean Baptiste, 68
Land bridges, 30–33
Lava, 20, 65ff. See also Volcanoes
Leakey, Louis, 102
Levine, Seymour, 49
Lights, earthquake, 35, 45–47, 48
Lister, Joseph, 79–80, 84, 86
Lomnitz, Cinna and Larissa, 46, 49

Magma, 19–20, 30
Malaria, 82–83, 85
Mammoth, woolly, 16–17
Martin, Robert, 102
Maui Island, 27, 30. See also Haleakala
Mauna Loa (volcano), 23, 27
Mazama, Mount, 11–14, 15
Medea legend, 74
Medicine (health), 62, 71–95
Mediterranean Sea, navigation of, 59–60
Meteorology, 63
Mice: and earthquakes, 49; and ginseng, 94
Minsky, Marvin, 114
Missouri earthquake, 36–41
Molds, 82, 86–89
Molokai Island, 21, 23, 27
Monkeys, behavior studies of, 104–5, 110
Moon, and weather, 64
Mormon tea bush, 77
Mountains. See Volcanoes

Neanderthals, 71, 72
Nevada, and earthquakes, 36

Newton, Isaac, 110, 115
Noah, 53
North Star, 58, 59
November, 59–60

Oahu Island, 21, 23, 27
Obsidian, 20
Odysseus, 58
Orion (constellation), 59, 61

Pacific Ocean, 54–58
Pacific plate, 30, 41
Paintings, cave, 16–17
Panhandling, study of, 107
Pasteur, Louis, 84, 88
Patterson, Penny, 106
Peking, China, and earthquake, 44
Pele, 21–23, 27
Pele's Hill, 21, 23
Penicillin, 87–89
Pentagon, 118–19
Phantom Ship (island), 15
Phenol, 79–80
Piezoelectric effect, 47
Plants, medicinal, 71–81, 93–94. *See also* specific plants
Plates, earth's 27–33, 41
Pleiades (stars), 54–56, 58, 59–60
Polaris (star), 58, 59
Polynesians, 54–58, 59. *See also* Hawaiian Islands

Quartz, 47

Ravens, 53
Rice, and beriberi, 90–91
Robots, 113–14
Romans, ancient, and stars, 58, 62

Salicylic acid, 79–82
San Andreas fault, 36, 40ff.
San Francisco earthquake, 36
Santa Rosa, California, earthquake, 45
Schaller, George, 97–98

Seismographs, 36, 39
Shanidar, Iraq, 71, 72
Shasta, Mount, 11, 14
Shetland Islands, 53
Simon, Herbert A., 110
Sirius (star), 61, 62
Sky, 53–70; earthquake lights, 35, 45–47, 48
Skye, Isle of, 19–21
Smith, Bruce, 49
Snow, 65
Solecki, Ralph S., 73
Stanford University, and earthquake studies, 49
Stars, 54–62
Sun: and volcano legend, 19ff.; and weather, 63–64

Takashi, Kamekiro, 90
Tangshan, China, earthquakes, 35, 44ff.
Taurus (constellation), 54, 55
Thiersch, Karl, 80
Tohoku University, Japan, and earthquake studies, 50
Tools and weapons, 17, 71, 98–99, 100
*2001* (movie), 100, 101
Tyndall, John, 86–87

Vietnam, 108–9, 119
Vitamins, 91–92
Volcanoes, 11–14, 19–33, 65
Voyager spacecraft, 119–20

Washoe (chimpanzee), 105, 106
Watanabe, Masao, 110
Weapons. *See* Tools and weapons
Weather, 61ff.
Weisskopf, Victor, 111
Weizenbaum, Joseph, 116–17
Wiener, Norbert, 117ff.
Willow bark, 79, 81
Wisconsin Primate Laboratory, 104–5
Witches, 86

Yasui, Yutaka, 46

Zulus, and robot legend, 113